# ANALOG ELECTRONIC DEVICES

## Theory and Practical

# ANALOG ELECTRONIC DEVICES
## Theory and Practical


AUTHOR

**Prof. N. B Balamurugan**
*Professor,*
Department of Electronics and Communication Engineering,
Thiagarajar College of Engineering, Madurai


---


REVIEWER

**Dr. Ankesh Jain**
*Assistant Professor,*
Department of Electrical Engineering,
IIT, Delhi



KHANNA BOOK PUBLISHING CO. (P) Ltd.
4C/4344, Ansari Road, Darya Ganj,
New Delhi, India

All India Council for Technical Education
Nelson Mandela Marg, Vasant Kunj,
New Delhi, India

Dear Readers,

To prevent the piracy, this book is secured with HIGH SECURITY HOLOGRAM on the front title cover. In case you don't find the hologram on the front cover title, please write us to at contact@khannabooks.com or whatsapp us at +91-99109 09320 and avail special gift voucher for yourself.

Specimen of Hologram on front Cover title:

Moreover, there is a SPECIAL DISCOUNT COUPON for you with EVERY HOLOGRAM.

How to avail this SPECIAL DISCOUNT:

Step 1: Scratch the hologram
Step 2: Under the scratch area, your "coupon code" is available
Step 3: Logon to www.khannabooks.com
Step 4: Use your "coupon code" in the shopping cart and get your copy at a special discount
Step 5: Enjoy your reading!

Thanks,
Khanna Publishing House

**ISBN:** 978-93-5538-380-8

**Book Code:** UG412EN

**ANALOG ELECTRONIC DEVICES**
Theory and Practical
*by* **Prof. N. B Balamurugan**

**First Edition:** 2026

To view complete list of books, please scan the QR Code:

Published by:
**Khanna Book Publishing Co. (P) Ltd.**
CIN: U22110DL1998PTC095547

Visit us at: www.khannabooks.com
Write us at: contact@khannabooks.com

Printed in Bharat

प्रो. टी. जी. सीताराम
अध्यक्ष
**Prof. T. G. Sitharam**
**Chairman**

सत्यमेव जयते

आजादी का अमृत महोत्सव

**अखिल भारतीय तकनीकी शिक्षा परिषद्**
(भारत सरकार का एक सांविधिक निकाय)
(शिक्षा मंत्रालय, भारत सरकार)
नेल्सन मंडेला मार्ग, वसंत कुंज, नई दिल्ली-110070
दूरभाष : 011-26131498
ई-मेल : chairman@aicte-india.org

**ALL INDIA COUNCIL FOR TECHNICAL EDUCATION**
(A STATUTORY BODY OF THE GOVT. OF INDIA)
(Ministry of Education, Govt. of India)
Nelson Mandela Marg, Vasant Kunj, New Delhi-110070
Phone : 011-26131498
E-mail : chairman@aicte-india.org

# FOREWORD

Engineers are the backbone of any modern society. They are the ones responsible for the marvels as well as the improved quality of life across the world. Engineers have driven humanity towards greater heights in a more evolved and unprecedented manner.

The All India Council for Technical Education (AICTE), have spared no efforts towards the strengthening of the technical education in the country. AICTE is always committed towards promoting quality Technical Education to make India a modern developed nation emphasizing on the overall welfare of mankind.

An array of initiatives has been taken by AICTE in last decade which have been accelerated now by the National Education Policy (NEP) 2020. The implementation of NEP under the visionary leadership of Hon'ble Prime Minister of India envisages the provision for education in regional languages to all, thereby ensuring that every graduate becomes competent enough and is in a position to contribute towards the national growth and development through innovation & entrepreneurship.

One of the spheres where AICTE had been relentlessly working since past couple of years is providing high quality original technical contents at Under Graduate & Diploma level prepared and translated by eminent educators in various Indian languages to its aspirants. For students pursuing 2nd year of their Engineering education, AICTE has identified 88 books, which shall be translated into 12 Indian languages - Hindi, Tamil, Gujarati, Odia, Bengali, Kannada, Urdu, Punjabi, Telugu, Marathi, Assamese & Malayalam. In addition to the English medium, books in different Indian Languages are going to support the students to understand the concepts in their respective mother tongue.

On behalf of AICTE, I express sincere gratitude to all distinguished authors, reviewers and translators from the renowned institutions of high repute for their admirable contribution in a record span of time.

AICTE is confident that these outcomes based original contents shall help aspirants to master the subject with comprehension and greater ease.

T.G. Sitharam

(Prof. T. G. Sitharam)

# Preface

The book titled "Analog Electronic Devices: Theory and Practicals is a result of our extensive expertise teaching fundamental semiconductor courses. The goal of developing this book is to introduce engineering students to basic principle of semiconductor devices, physics basics, and provide them with an understanding of the topic. We have incorporated the themes specified by AICTE in a highly methodical and organised manner throughout the book, keeping in mind the objective of broad coverage as well as providing necessary extra material. Attempts have been made to convey the essential principles of the subject as simply as feasible. The objective of this book is to combine quantum mechanics, quantum theory of solids, semiconductor material physics, and semiconductor device physics together. The quantity of physics presented in this text is greater than what is covered in many introductory semiconductor device books. Although this topic is deeper, the author has discovered that after thoroughly addressing the fundamental introduction and material physics, the physics of the semiconductor device follows very easily and can be studied pretty fast and efficiently. The attention on the underlying physics will also aid understanding and maybe the development of novel semiconductor devices.

During the process of preparation of the manuscript, we have considered the various standard text books and accordingly we have developed sections like critical questions, solved and supplementary problems etc. While preparing the different sections emphasis has also been laid on definitions and laws and also on comprehensive synopsis of formulae for a quick revision of the basic principles. The book covers all types of medium and advanced level problems and these have been presented in a very logical and systematic manner. The gradations of those problems have been tested over many years of teaching to a wide variety of students.

Apart from illustrations and examples as required, we have enriched the book with numerous solved problems in every unit for proper understanding of the related topics. Under the common title "Analog Electronic Devices: Theory and Practicals" there is a set of four books covering different aspects and applications of physics in engineering. Out of those, the first one covers Introduction to semiconductor physics, the second one is based on semiconductor diode, the third one is related to application diode for Engineers and the fourth one is based on Bipolar Junction Transistor, field effect transistor, Integrated circuit fabrication. It is important to note that in all the books, we have included the relevant laboratory practical. In addition, besides some essential information for the users under the heading "Know More" we have clarified some essential basic information in the appendix and annexure section.

As far as the present book is concerned, "Analog Electronic Devices: Theory and Practicals" is meant to provide a thorough grounding in applied physics on the topics covered. This part of the electronics book will prepare engineering students to apply the knowledge of physics to tackle 21st century and onward engineering challenges and address the related aroused questions. The subject matters are presented in a constructive manner so that an Engineering degree prepares students to work in different sectors or in national laboratories at the very forefront of technology.

We sincerely hope that the book will inspire the students to learn and discuss the ideas behind basic principles of engineering physics and will surely contribute to the development of a solid foundation of the subject. We would be thankful to all beneficial comments and suggestions which will contribute to the improvement of the future editions of the book. It gives us immense pleasure to place this book in the hands of the teachers and students. It was indeed a big pleasure to work on different aspects covering in the book.

**Prof. N B Balamurugan**

# Acknowledgement

The authors are grateful to the authorities of AICTE, particularly Prof. T. G. Sitharam, Chairman; Dr. Abhay Jere, Vice-Chairman; Prof. Rajive Kumar, Member-Secretary;

Dr. Ramesh Unnikrishnan, Advisor-II and Dr. Sunil Luthra, Director, Training and Learning Bureau, for their planning to publish the books on Analog Electronic Devices: Theory and Practicals. We sincerely acknowledge the valuable contributions of the reviewer of the book Prof. Ankesh Jain, IIT Delhi, for making it students' friendly and giving a better shape in an artistic manner.

This book is an outcome of various suggestions of AICTE members, experts and authors who shared their opinion and thought to further develop the engineering education in our country. Acknowledgements are due to the contributors and different workers in this field whose published books, review articles, papers, photographs, footnotes, references and other valuable information enriched us at the time of writing the book.

I convey heartfelt thanks to the PhD scholars M Hemalatha, E Rajalakshmi and S Rajkumar for their time, expertise, and dedication to complete this book within the given time. Last but not the least, the book would not have been possible without the support and patience of my family. They have borne my addiction to the book very very patiently.

**Prof. N B Balamurugan**

# Outcome Based Education

For the implementation of an outcome based education the first requirement is to develop an outcome based curriculum and incorporate an outcome based assessment in the education system. By going through outcome based assessments evaluators will be able to evaluate whether the students have achieved the outlined standard, specific and measurable outcomes. With the proper incorporation of outcome based education there will be a definite commitment to achieve a minimum standard for all learners without giving up at any level. At the end of the programme running with the aid of outcome based education, a student will be able to arrive at the following outcomes:

**PO1.** **Engineering knowledge:** Apply the knowledge of mathematics, science, engineering fundamentals, and an engineering specialization to the solution of complex engineering problems.

**PO2.** **Problem analysis:** Identify, formulate, review research literature, and analyze complex engineering problems reaching substantiated conclusions using first principles of mathematics, natural sciences, and engineering sciences.

**PO3.** **Design / development of solutions:** Design solutions for complex engineering problems and design system components or processes that meet the specified needs with appropriate consideration for the public health and safety, and the cultural, societal, and environmental considerations.

**PO4.** **Conduct investigations of complex problems:** Use research-based knowledge and research methods including design of experiments, analysis and interpretation of data, and synthesis of the information to provide valid conclusions.

**PO5.** **Modern tool usage:** Create, select, and apply appropriate techniques, resources, and modern engineering and IT tools including prediction and modeling to complex engineering activities with an understanding of the limitations.

**PO6.** **The engineer and society:** Apply reasoning informed by the contextual knowledge to assess societal, health, safety, legal and cultural issues and the consequent responsibilities relevant to the professional engineering practice.

**PO7.** **Environment and sustainability:** Understand the impact of the professional engineering solutions in societal and environmental contexts, and demonstrate the knowledge of, and need for sustainable development.

**PO8.** **Ethics:** Apply ethical principles and commit to professional ethics and responsibilities and norms of the engineering practice.

**PO9.** **Individual and team work:** Function effectively as an individual, and as a member or leader in diverse teams, and in multidisciplinary settings.

**PO10. Communication:** Communicate effectively on complex engineering activities with the engineering community and with society at large, such as, being able to comprehend and write effective reports and design documentation, make effective presentations, and give and receive clear instructions.

**PO11. Project management and finance:** Demonstrate knowledge and understanding of the engineering and management principles and apply these to one's own work, as a member and leader in a team, to manage projects and in multidisciplinary environments.

**PO12. Life-long learning:** Recognize the need for, and have the preparation and ability to engage in independent and life-long learning in the broadest context of technological change.

# Course Outcomes

After completion of the course the students will be able to:

**CO-1:** Describe the principles of semiconductor Physics.

**CO-2:** Identify passive and active components based on their characteristics.

**CO-3:** Apply the knowledge of diodes and special purpose diodes in various applications.

**CO-4:** Distinguish between BJT and FET.

**CO-5:** Design and analyze BJT and FET amplifiers.

**CO-6:** Understand the integrated circuit fabrication process.

| Course Outcomes | Expected Mapping with Programme Outcomes (1- Weak Correlation; 2- Medium correlation; 3- Strong Correlation) | | | | | | | | | | | |
|---|---|---|---|---|---|---|---|---|---|---|---|---|
| | PO-1 | PO-2 | PO-3 | PO-4 | PO-5 | PO-6 | PO-7 | PO-8 | PO-9 | PO-10 | PO-11 | PO-12 |
| CO-1 | 2 | 1 | - | - | 1 | - | - | - | - | - | - | - |
| CO-2 | 2 | 1 | - | - | 1 | - | - | - | - | - | - | - |
| CO-3 | 3 | 2 | 1 | - | 1 | - | - | - | - | - | - | - |
| CO-4 | 3 | 2 | 1 | - | 1 | - | - | - | - | - | - | - |
| CO-5 | 3 | 2 | 1 | - | 1 | - | - | - | 1 | 1 | - | - |
| CO-6 | 2 | 1 | - | - | - | - | - | - | 1 | - | - | - |

# Guidelines for Teachers

To implement Outcome Based Education (OBE) knowledge level and skill set of the students should be enhanced. Teachers should take a major responsibility for the proper implementation of OBE. Some of the responsibilities (not limited to) for the teachers in OBE system may be as follows:

- Within reasonable constraint, they should manoeuvre time to the best advantage of all students.
- They should assess the students only upon certain defined criterion without considering any other potential ineligibility to discriminate them.
- They should try to grow the learning abilities of the students to a certain level before they leave the institute.
- They should try to ensure that all the students are equipped with the quality knowledge as well as competence after they finish their education.
- They should always encourage the students to develop their ultimate performance capabilities.
- They should facilitate and encourage group work and team work to consolidate newer approach.
- They should follow Blooms taxonomy in every part of the assessment.

## Bloom's Taxonomy

| Level | Teacher should Check | Student should be able to | Possible Mode of Assessment |
|---|---|---|---|
| Create | Students ability to create | Design or Create | Mini project |
| Evaluate | Students ability to justify | Argue or Defend | Assignment |
| Analyse | Students ability to distinguish | Differentiate or Distinguish | Project/Lab Methodology |
| Apply | Students ability to use information | Operate or Demonstrate | Technical Presentation/ Demonstration |
| Understand | Students ability to explain the ideas | Explain or Classify | Presentation/Seminar |
| Remember | Students ability to recall (or remember) | Define or Recall | Quiz |

## Guidelines for Students

Students should take equal responsibility for implementing the OBE. Some of the responsibilities (not limited to) for the students in OBE system are as follows:

- Students should be well aware of each UO before the start of a unit in each and every course.
- Students should be well aware of each CO before the start of the course.
- Students should be well aware of each PO before the start of the programme.
- Students should think critically and reasonably with proper reflection and action.
- Learning of the students should be connected and integrated with practical and real-life consequences.
- Students should be well aware of their competency at every level of OBE.

# Symbols and Abbreviations

## LIST OF ABBREVIATIONS

| General Terms | | | |
|---|---|---|---|
| **Abbreviations** | **Full form** | **Abbreviations** | **Full form** |
| E-K Diagram | Energy Momentum Diagram | GSI | Giga Scale Integration |
| DIL | Dual-In-Line | CMOS | Complementary Metal Oxide Semiconductor |
| LED | Light Emitting Diode | NMOS | N-type Metal Oxide Semiconductor |
| OP-AMP | Operational Amplifier | PMOS | P-type Metal Oxide Semiconductor |
| TTL | Transistor-Transistor Logic | MGS | Metallurgical Grade Silicon |
| PV | Photo Voltaic | EGS | Electronic Grade Silicon |
| KCL | Kirchhoff's Current Law | SGS | Semiconductor Grade Silicon |
| CFL | Compact Fluorescent Lamp | CVD | Chemical Vapour deposition |
| LCD | Liquid Crystal Display | APCVD | Atmospheric-pressure Chemical Vapour Deposition |
| CRI | Colour Rendering Index | LPCVD | Low Pressure Chemical Vapour Deposition |
| IR | Reverse Current | PECVD | Plasma enhanced Chemical Vapour Deposition |
| BJT | Bipolar Junction Transistor | VPE | Vapour Phase Epitaxy |
| Q point | Quiescent point | PVD | Physical Vapour Deposition |
| KVL | Kirchhoff's Voltage Law | SMT | Surface Mount Technology |
| S | Stability factor | PCB | Printed Circuit Board |
| FET | Field Effect Transistor | SIP or SIPP | Single in-line Pin Package |
| JFET | Junction Field Effect Transistor | DIP | Dual In-line package |
| MOSFET | Metal Oxide Semiconductor Field Effect Transistor | ZIP | Zig-Zag in -Line Package |
| E-MOSFET | Enhancement MOSFET | SOIC | Small Out line Integrated Circuit |
| D-MOSFET | Depletion MOSFET | SOP | Small Outline Package |
| MOS | Metal Oxide Semiconductor | QFP | Quad Flat Package |
| $Al_2O_3$ | Aluminum oxide | PLCC | Plastic Leaded Chip Carrier |
| $SiO_2$ | Silicon dioxide | MIC | Monolithic IC |

| IGFET | Insulated Gate Field Effect transistor | HIC | Hybrid IC |
|---|---|---|---|
| IC | Integrated Circuits | SSI | Small Scale Integration |
| Si | Silicon | MSI | Medium Scale Integration |
| Ge | Germanium | LSI | Large Scale Integration |
| GaAs | Gallium Arsenide | VLSI | Very Large-Scale Integration |
| ULSI | Ultra Large-Scale Integration | BGA | Ball Grid Array |
| HF | Hydrofluoric Acid | | |

## List of Symbols

| Symbols | Description | Symbols | Description |
|---|---|---|---|
| h | Plank's Constant | $W$ | width |
| $\phi$ | Material work function | $G_{no}$ | Electron thermal generation rate |
| p | Momentum of the particle | $G_{po}$ | Hole thermal generation rate |
| $\lambda$ | Wavelength of the particle | $R_{no}$ | Recombination of electrons |
| m | Mass of the particle | $R_{po}$ | Recombination of holes |
| v | Velocity | $n_o$ | Electron concentration at thermal equilibrium |
| c | Speed of light in vacuum | $p_o$ | Hole concentration at thermal equilibrium |
| $\psi(x,t)$ | Wave function | $\delta n$ | Excess electron concentration |
| $V(x)$ | Potential function | $\delta p$ | Excess hole concentration |
| $\eta$ | Separation Constant | $g'_n$ | Excess rate of electron generation |
| E | Particle energy | $g'_p$ | Excess rate of hole generation |
| $E_G$ | Forbidden energy gap (or) Band gap | $R'_n$ | Excess rate of electron recombination |
| $e_V$ | Electron volt | $R'_p$ | Excess rate of hole recombination |
| $E_F$ | Fermi Level | $\tau_{no}$ | Excess lifetime of electron |
| $E_C$ | Lowest conduction band energy level | $\tau_{po}$ | Excess lifetime of hole |
| $E_V$ | Highest valence band energy level | $n_i$ | Intrinsic carrier concentration |
| Si | Silicon | $GaAs$ | Gallium Arsenide |
| $G_e$ | Germanium | $D$ | Electric displacement field |
| $A_s$ | Arsenic | $\hat{\epsilon}$ | Permittivity tensor |
| $E_D$ | Donor energy level | $N_A^-$ | Ionized Acceptor |
| B | Boron | $N_D^+$ | Ionized Donor |
| $E_{Fi}$ | Intrinsic fermi level | $N_{ox}$ | Oxide charges concentration |

| | | | |
|---|---|---|---|
| $E_A$ | Acceptor energy level | $N_{it}$ | Interface trap charge concentration |
| K | Momentum | $N_A$ | Concentration of acceptor impurity |
| $V_n$ | Drift velocity of electron | $N_D$ | Concentration of donor impurity |
| $V_p$ | Drift velocity of holes | $k$ | Boltzmann's constant |
| $\mu_n$ | Mobility of electrons | $T$ | Room temperature |
| $\mu_p$ | Mobility of holes | $V_{bi}$ | Built in potential |
| $J_{n,drift}$ | Drift current density of electron | $V_D$ | Diode voltage |
| $J_{p,drift}$ | Drift current density of hole | $I_D$ | Diode current |
| e | Charge of an electron | $V_z$ | Breakdown voltage or Zener voltage |
| $J_d$ | Overall drift current density | $I_z$ | Zener current |
| n | Concentration of electrons | $\chi$ | Electron affinity |
| p | Concentration of holes | $\phi_m$ | Metal workfunction |
| $J_{n,diff}$ | Diffusion current density of electron | $\phi_s$ | Semiconductor workfunction |
| $J_{p,diff}$ | Diffusion current density of hole | $I_E$ | Emitter current |
| J | Total current density | $I_c$ | Collector current |
| $V_d$ | Drift velocity | $I_B$ | Base Current |
| E | Electric field | $\propto_F$ | Common base current gain |
| a | Acceleration of carriers | $I_{co}$ | Reverse bias base-collector junction current |
| $\tau$ | Mean collusion time | $I_R$ | Reverse current |
| $\rho$ | Resistivity | $R_{IN}$ | Input Resistance |
| L | Length | $r_d$ | Drain resistance |
| A | Area | $\mu$ | Amplification factor |
| R | Resistance | $P_D$ | Power dissipation |
| $\sigma$ | Conductivity | $I_G$ | Gate current |
| t | Thickness | $t_{ox}$ | Oxide thickness |
| $R_s$ | Sheet resistance | $\varepsilon_{ox}$ | Permittivity of oxide |
| $I_F$ | Forward current | $Q$ | Charge |
| $I_{Eo}$ | Reverse bias base-emitter junction current | $\psi_s$ | Surface potential |
| $I_{CBo}$ | Reverse saturation current | $V_T$ | Threshold voltage |
| $W_B$ | Base width | $K$ | Conduction Parameter |
| $V_{BE}$ | Base-Emitter voltage | $I_{DQ}$ | Quiescent drain current |
| $V_{CE}$ | Collector-Emitter voltage | $\lambda$ | Channel length modulation parameter |
| $\beta$ | Common Emitter amplification factor | $r_o$ | Output resistance |

| $\gamma$ | Emitter injection efficiency | $O_2$ | Oxygen gas |
|---|---|---|---|
| SiC | Silicon Carbide | $H_2O$ | Water vapor |
| $V_{GS}$ | Gate-source voltage | $H_2$ | Hydrogen gas |
| $V_{DS}$ | Drain-Source voltage | $Na$ | Sodium |
| $I_D$ | Drain current | $B_2O_3$ | Boron Oxide |
| $V_p$ | Pinch-off voltage | $BCL_3$ | Boron chloride |
| $V_{SG}$ | Source-Gate voltage | $P_2O_3$ | Phosphorous Oxide |
| $g_m$ | Transconductance or mutual conductance | $POCL_3$ | Phosphorous oxychloride |

# List of Figures

## Unit 1: Semiconductor Physics



## Unit 2: Semiconductor Diodes

## Unit 3: Application of Diode

## Unit 4: Bipolar Junction Transistor

## Unit 5: Field Effect Transistor

## Unit 6: Integrated Circuit Fabrication Process

# Contents

# 01 SEMICONDUCTOR PHYSICS

## UNIT SPECIFICS

*Through this unit we have discussed the following aspects:*

- *Quantum mechanics fundamentals that relate to semiconductor physics*
- *Fundamental properties of electron behaviour in periodic lattice*
- *Relation between energy and momentum and application of E-K diagram*
- *The technique by which impurities are added to semiconductors to change their characteristics.*
- *The mechanism of drift and diffusion current on the application of field.*
- *The characteristics of carrier mobility*
- *Design of resistors and sheet resistance*

*The practical applications of the topics are discussed for generating further curiosity and creativity as well as improving problem solving capacity.*

*Besides giving a large number of multiple-choice questions as well as questions of short and long answer types marked in two categories following lower and higher order of Bloom's taxonomy, assignments through a number of numerical problems, a list of references and suggested readings are given in the unit so that one can go through them for practice. It is important to note that for getting more information on various topics of interest some QR codes have been provided in different sections which can be scanned for relevant supportive knowledge.*

*After the related practical, based on the content, there is a "Know More" section. This section has been carefully designed so that the supplementary information provided in this part becomes beneficial for the users of the book. This section mainly highlights the initial activity, examples of some interesting facts, analogy, history of the development of the subject focusing the salient observations and finding, timelines starting from the development of the concerned topics up to the recent time, applications of the subject matter for our day-to-day real life or/and industrial applications on variety of aspects, case study related to environmental, sustainability, social and ethical issues whichever applicable, and finally inquisitiveness and curiosity topics of the unit.*

## RATIONALE

*This chapter is intended to equip students in electronic material and devices with the fundamentals of semiconductor physics. The materials covered in the module begins with fundamentals and accelerates to advanced topics in semiconductor physics. All the basic aspects are relevant to basic concepts necessary to understand the fundamentals of semiconductor devices. It starts with quantum physics extended to semiconductor mechanism. It then explains clearly the semiconductor material and carrier transport mechanism of electron and holes in semiconductor. Understanding these charged particle concentrations is crucial for interpreting an electrical characteristics of a semiconductor materials. The physical characteristics of the semiconductor device are covered efficiently. The focus on physical mechanism will also aid understanding and development of novel semiconductor technologies.*

## PRE-REQUISITES

*Fundamentals of Basic Physics*

## UNIT OUTCOMES

***List of outcomes of this unit is as follows:***

*U1-O1: Describe about basic quantum mechanics*
*U1-O2: Analyses of E-K diagram of semiconductor*
*U1-O3: Describe the fundamentals of semiconductor materials*
*U1-O4: Realize carrier transport mechanism in a semiconductor*
*U1-O5: Design of resistor and sheet resistance*

| Unit-1 Outcomes | EXPECTED MAPPING WITH COURSE OUTCOMES (1- Weak Correlation; 2- Medium Correlation; 3- Strong Correlation) | | | | | |
|---|---|---|---|---|---|---|
| | CO-1 | CO-2 | CO-3 | CO-4 | CO-5 | CO-6 |
| U1-01 | 3 | 3 | 3 | - | 3 | 1 |
| U1-02 | 1 | 1 | 2 | 2 | 1 | - |
| U1-03 | 2 | 1 | 3 | 1 | 2 | 1 |
| U1-04 | - | - | 3 | 1 | 2 | 2 |
| U1-05 | 3 | 3 | 3 | - | 3 | 1 |

## 1.1 QUANTUM MECHANICS

In the first three-quarters of this century, it was found that a large variety of phenomena involving electrons, atoms, and light could not be well explained by the laws of motion developed by Galileo, Newton, Lagrange, Hamilton, and Maxwell, among many others. After much research, a new hypothesis (along with a novel pattern of motion) was revealed in 1924. This concept is called "Quantum Mechanics", which serves as the foundation for comprehending atomic, nuclear, sub-atomic, condensed matter, and "Solid-state" physics. There are three concepts to examine before diving into the mathematics of quantum mechanics: the energy quanta principle, the De-Broglie hypothesis, and the Heisenberg uncertainty principle.

### 1.1.1 Energy Quanta

The discrepancy among the experimental data and the conventional light theory are explained by the term "Photoelectric Effect". When monochromatic light falls on a metal surface of a material, this incident light is absorbed by the metal electrons, and it comes out from the surface. These electrons are known as photoelectrons. As per the classical theory, when the intensity of the light is high enough, we would expect greater number of electrons to be expelled from the surface regardless of the incoming frequency. The observed result shows that the number of electrons that will emit from the metal surface does not depend on the intensity of light, it mainly depends on the frequency ($\nu=\nu_0$) of light. Below this frequency no photoelectron is created at a stable intensity of the event.

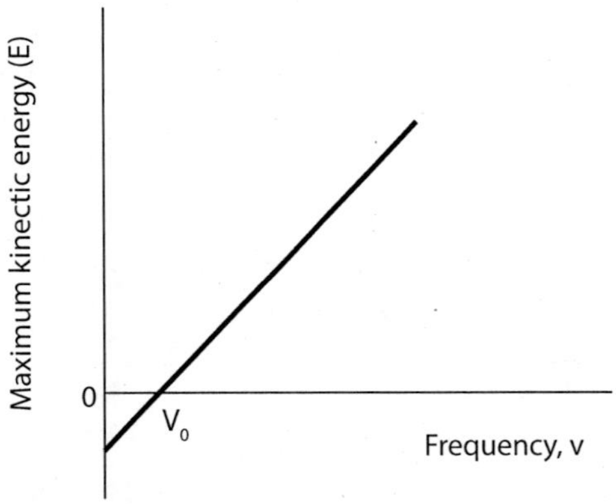


**Fig.1.1:** Maximum Kinetic energy as function of incident energy

Fig. 1.1 shows that at a constant incident intensity, the maximum kinetic energy of the photoelectrons varies linearly with frequency. The rate of photoelectron emission changes when the incoming intensity fluctuates with a fixed frequency, while the maximal kinetic energy remains constant. In 1990, Planck proposed that the radiation emitted from the heating surface is not continuous, they come in discrete packets known as "quanta". The quanta's energy (E) and frequency ν are related by $E = h\nu$. Where ν is the radiation frequency and is the Plank's constant with the constant value ($h = 6.625 \times 10^{-34}$ *J.s*). This mechanism is known as "Planck's Radiation Law". Later, in 1905, Einstein came forward and considered the particle nature of the light known as "*photon*".

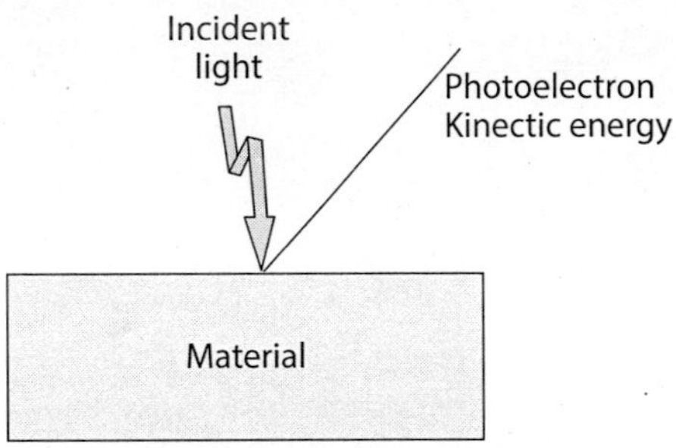


**Fig.1.2:** Photoelectric Effect

The photoelectric effect illustrates the distinct nature of the photon as well as its particle like behaviour. A photon with enough energy can knock an electron from the surface of the material. The minimum amount of energy needed to remove an electron is known as work function of the material, as well as any extra energy created by a photon is converted to kinetic energy as shown in Fig. 1.2.

The maximum photoelectron kinetic energy ($KE_{\_max}$) is given by,

$$KE_{\_max} = 1/2\, mv^2 = h\nu - \phi = h\nu - h\nu_0, (\nu \geq \nu_0) \quad (1.1)$$

where $h\nu$ is the incident photon energy and $\phi = h\nu_0$ is material's work function or minimum energy.

### 1.1.2 De-Broglie Hypothesis

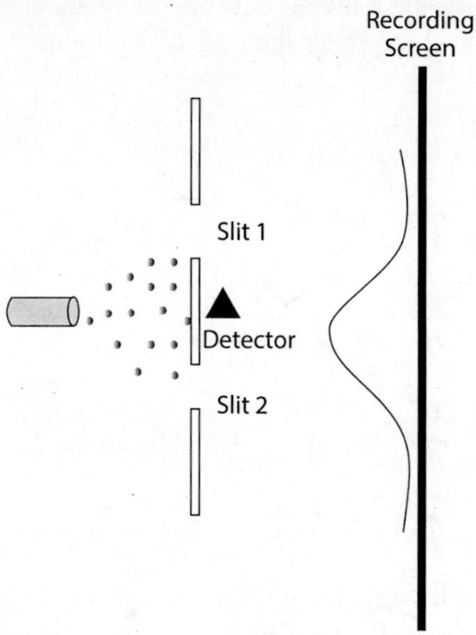


**Fig.1.3:** Feynman's Famous Thought Experiment

The better way to explain quantum mechanics is the famous thought experiment by Richard Feynman. We are going to play around with this experiment and try to understand the behaviour of a particle.

The experimental setup of Feynman's experiment is shown in Fig. 1.3. The setup consists of an electron gun. This electron gun essentially provides tiny discrete particles. The detector is placed near the slits to detect the path of the particle (i.e., slit 1 or slit 2). The recording screen is placed to detect the particle which strikes that screen. The experiment is telling that despite the electrons being fired one at a time with some sufficient space and time between firing, each electron has gone through both slits simultaneously much like a "*wave*" not like a "*particle*".

This wave-like behaviour of the particle is defined by De-Broglie Hypothesis. In 1924, he postulated that particle exhibits wave-like behaviour. This is known as "*principle of duality (Wave-particle)*". According to the hypothesis, the photon momentum is expressed as,

$$p = \frac{h}{\lambda} \tag{1.2}$$

Then, wavelength of the particle can be expressed as,

$$\lambda = \frac{h}{p} \tag{1.3}$$

The momentum of particle ($p$) and it is expresses as $p = mv$,

where is mass of the particle, is velocity of the particle and is known as De-Broglie wavelength.

**Example 1.1:** Consider the particle travelling with the velocity of , then calculate the wavelength of the particle. (Hint: use De-Broglie wavelength)

**Given:** Velocity, $v = 10^{10}$ $cm/s = 10^8$ m/s

**Solution:**

We know that mass of the particle $m = 9.1\times10^{-31}$ kg

From the momentum equation of the particle, we can calculate the momentum,

$$p = mv = 9.1 \times 10^{-31} \times 10^{8} = 9.1 \times 10^{-23} \text{ kg. } m/s$$

Then the wavelength is calculated by,

$$\lambda = \frac{h}{p} = \frac{6.625 \times 10^{-34}}{9.1 \times 10^{-23}} = 0.72 \times 10^{-11} \text{ m}$$

## 1.1.3 Heisenberg Uncertainty Principle

The uncertainty principle explains how two conjugate variables such as position and momentum, as well as energy and time, are related. In 1927, Heisenberg applied this principle to sub-atomic particles which are very small in size and it claims that their behaviour is unpredictable. So, it is not possible to measure the position and momentum absolutely.

The first statement in the uncertainty principle describes that, absolute precision in both position and momentum of the particle are not attainable simultaneously. Let us consider is the uncertainty of momentum and the uncertainty in the position is represented by the parameter . Then the principle is defined by,

$$\Delta p \Delta x = \frac{h}{2} \tag{1.4}$$

The second statement is that, it is difficult to accurately characterize a particle's energy (E) and the instant time at which the particle has this same energy. So, and represent the uncertainty in energy and time respectively. Then the principle is defined as,

$$\Delta E \Delta t \geq \frac{\hbar}{2} \tag{1.5}$$

where $\hbar = \frac{h}{2} = 1.054 \times 10^{(-34)}$ $J.s$ and it is known as modified Planck's constant.

## 1.1.4 Schrodinger's Wave Equation

It became clear that mechanics needed to be reconstructed as a consequence of several experimental findings involving electromagnetic waves and particles that could not be described by conventional physics. In 1926, Schrodinger proposed a wave mechanics formulation that combined Planck's quanta principles with De-Broglie's wave-particle duality concept. In order to explain the movement of electrons in a crystal, we can utilize wave theory, based on the wave-particle duality concept. Schrodinger's wave equation describes this wave theory. The one-dimensional wave equation is given by,

Potential energy + Kinetic energy = Total energy

$$-\frac{\hbar}{2\text{m}} \frac{\partial^2 \psi(x,t)}{\partial x^2} + \text{V}(x)\psi(x,t) = -\frac{\hbar}{j} \frac{\partial \psi(x,t)}{\partial t} \tag{1.6}$$

where,

wave function - $\psi(x,t)$,

potential function - $V(x)$, considered to be independent of time,

$J$ (imaginary constant) is defined as $\sqrt{-1}$,

$m$ -Mass of the particle.

The Schrodinger's equation can be separated into time-dependent component of wave function and time-independent component of the wave function. The time-dependent component of the wave function and the position-dependent component of the wave equation can be determined by using the technique called separation of variables. Assume that the wave function can be written in the form,

$$\psi(x,t) = \psi(x)\phi(t) \tag{1.7}$$

where $\psi(x)$ is a function of the position x only and is a function of time t only. Substitute this to the equation 1.6,

$$-\frac{\hbar}{2m}\phi(t)\frac{\partial^2\psi(x)}{\partial x^2} V(x)\phi(t)\psi(x) = -\frac{\hbar}{j}\psi(x)\frac{\partial\phi(x,t)}{\partial t} \tag{1.8}$$

If we divide the above equation by the total wave function equation 1.7, the equation 1.8 becomes,

$$-\frac{\hbar}{2m}\frac{1}{\psi(x)}\frac{\partial^2\psi(x)}{\partial x^2} + V(x) = -\frac{\hbar}{j}\frac{1}{\psi(x)}\frac{\partial\phi(t)}{\partial t} \tag{1.9}$$

The left side of the equation 1.9 depends on a function of position x only while the right side of the equation is a function of time t only. The time-dependent wave function is defined as,

$$\eta = j\hbar\frac{1}{\psi(x)}\frac{\partial\phi(t)}{\partial t} \tag{1.10}$$

where $\eta$ is the separation constant and $\phi(t)$ is the function of time only. The solution to the equation 1.10 is written in the form of $\phi(t) = e^{-j(E/\hbar)t}$

The time-independent wave function is defined as,

(1.11)

We can denote the separation constant in terms of particle's energy (E). Multiply the term by $-\frac{2m}{\hbar^2}\psi(x)$ on both sides of the equation 1.11,

$$\frac{\partial^2\psi(x)}{\partial x^2} + \frac{2m}{\hbar^2}(E - V(x))\psi(x) = 0 \tag{1.12}$$

The wave function $\psi(x,t)$ describes the behaviour of an electron in a crystal. The total wave function is the product of position-dependent and time-dependent.

$$\psi(x,t) = \psi(x)\phi(t) = \psi(x)\, e^{-j(E/\hbar)t} \tag{1.13}$$

In 1926, Max Born postulated that the function $|\psi(x,t)|^2\, dx$ is the probability density function.

$$|\psi(x,t)|^2 = \psi(x,t).\psi^*(x,t) \tag{1.14}$$

where, is the complex conjugate function. Therefore, substituting this to the equation 1.13 gives,

$$|\psi(x,t)|^2 = \psi(x)\,\psi^*(x) = |\psi(x)|^2 \tag{1.15}$$

In classical mechanics, the position of the particle can be determined precisely; whereas in quantum mechanics, the position of the particle can be found in terms of probability.

Since the function $|\psi(x,t)|^2$ is the probability density function, then for a single particle,

$$\int_{-\infty}^{\infty} |\psi(x)|^2 \, dx = 1 \tag{1.16}$$

The above equation defines that the probability of finding the particle somewhere is certain and allows to normalize the wave function. This function is used to find the coefficients of some wave function.

## 1.2 PARTICLE IN A CRYSTAL LATTICE

The element in a one-dimensional lattice is an issue in quantum mechanics that appears in the periodic crystal lattice model. Electrons inside the lattice are subject to a consistent potential due to the electromagnetic field that is produced by the ions in the periodic framework of the crystal. The unbound electron model, which presumes that the lattice has no potential, is generalized in this model.

### *Problem Definition*

When discussing solid materials, periodic lattices in crystals are the principal topic. Let's discuss about a 1D lattice of positive ions here. Consider the distance between two ions is "a", then the potential in the lattice is shown in Fig. 1.4.

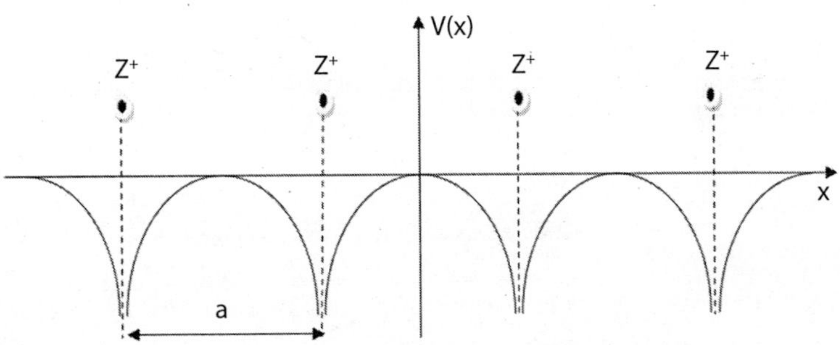


**Fig.1.4:** The representation of the particle in a potential well with a period 'a'

The potential is represented mathematically as a periodic function with period "a". When the potential is periodic, Bloch's theorem states that, the Schrodinger equation's wave function solution can be expressed as follows,

$$\psi(x)=e^{ikx}\, u(x) \tag{1.17}$$

where $u(x)$ is a periodic function which satisfies $u(x+a)=u(x)$.

For a periodic potential, the Bloch factor with the Floquet exponent k creates the band structure of the energy spectrum of the Schrodinger equation, such as the Kronig–Penney potential or a cosine function as in the Mathieu equation.

The boundary condition fails as one gets closer to the lattice's edges. If "L" is the length of the lattice such that L>>a, there are so many lattice ions. When analysing just one ion, its surroundings are just about linear and the electron wave function is unaffected. Instead of two circular boundary conditions, now it is having only one condition, as seen below.

$$\psi(0) = \psi(L) \tag{1.18}$$

The relationship aN=L exists if N is the number of ions in the lattice. Apply Bloch's theorem and replacement in the boundary condition led to a quantization for k.

$$\psi(0) = e^{ik0}\, u(0) = e^{ikL} u(L) = \psi(L)$$

$$u(0) = e^{ikL}\, u(L) = e^{ikL}\, u(aN) \rightarrow e^{ikL} = 1$$

$$kL = 2\pi n \rightarrow k = \frac{2\pi n}{L} \qquad \left(n = 0, \pm 1, \cdots, \pm \frac{N}{2}\right)$$

### 1.2.1 Model of Kronig-Penney

The Kronig Penney model is a straight forward, idealised quantum mechanical structure made up of a virtually infinite number of periodic arrays of rectangular potential barriers. A rectangular potential can be used to approximate the potential function is shown in Fig. 1.5.

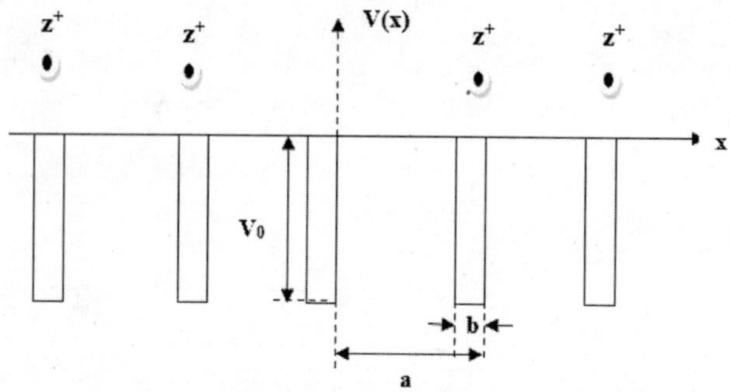


**Fig.1.5:** The rectangular finite potential well

Bloch's theorem tells that, to determine a continuous and smooth solution for a single time frame and the function u(x), which is also continuous and smooth.

A single instance of the potential is taken into consideration. Since there are two different regions, we will specifically find solutions for each of them. Consider E represent the energy above the well (E > 0).

**For 0 < x < (a-b):**

$$-\frac{\hbar^2}{2m}\psi_{xx} = E\psi$$

$$\psi = Ae^{i\alpha x} + A'e^{-i\alpha x}, \left(\alpha^2 = \frac{2mE}{h^2}\right)$$

Scan to know more

**For -b < x < 0:**

$$-\frac{\hbar^2}{2m}\psi_{xx} = (E+V_0)\psi$$

$$\psi = Be^{i\beta x} + BA'e^{-i\beta x}, \left(\beta^2 = \frac{2m(E+V_0)}{\hbar^2}\right)$$

In order to determine u(x) in each region, modify the electron wavefunction.

$\psi\,(0 < x < a - b) = Ae^{i\alpha x} + A'\,e^{-i\alpha x} = e^{ikx}\,(Ae^{i(\alpha-k)x} + A'\,e^{-i(\alpha+k)x})$

$u\,(0 < x < a - b) = Ae^{i(\alpha-k)x} + A'\,e^{-i(\alpha+k)x}$

In a similar way,

$u(-b < x < 0) = Be^{i(\beta-k)x} + B'\,e^{-i(\beta+k)x}$ (1.20)

Now ensure that the probability function is continuous and smooth in order to complete the solution, i.e.

$\psi(0^+) = \psi(0^-)$

$\psi'\,(0^+) = \psi'\,(0^-)$ (1.21)

Additionally, the periodicity of u(-x) and u'(-x) is given by,

$u(-b) = u(a-b)$

$u'\,(-b) = u'\,(a-b)$

These circumstances result in the matrix shown below:

$$\begin{pmatrix} 1 & 1 & -1 & -1 \\ \alpha & -\alpha & -\beta & \beta \\ e^{i(\alpha-k)(a-b)} & e^{-i(\alpha-k)(a-b)} & -e^{-i(\beta-k)b} & -e^{i(\beta+k)b} \\ (\alpha-k)e^{i(\alpha-k)(a-b)} & -(\alpha-k)e^{-i(\alpha-k)(a-b)} & -(\beta-k)e^{-i(\beta-k)b} & (\beta+k)e^{-i(\beta+k)b} \end{pmatrix}\begin{pmatrix} A \\ A' \\ B \\ B' \end{pmatrix} = \begin{pmatrix} 0 \\ 0 \\ 0 \\ 0 \end{pmatrix}$$

To get a non-trivial solution, determinant of the matrix needs to be zero. This results in the phrase that follows:

$$\cos(ka) = \cos(\beta b)\cos[(\alpha(a-b)] - \frac{(\alpha^2+\beta^2)}{2\alpha\beta}\sin(\beta b)\sin[(\alpha(a-b)]$$

Use the following approximations, to further reduce the expression,

$b \to 0;\; V_0 \to \infty;\; V_0 b = \text{constant}$

$\beta^2\, b = \text{constant};\; \alpha^2\, b \to 0$

$\beta b \to 0;\; \sin(\beta b) \to \beta b;\; \cos(\beta b) \to 1$

Now, the phrase will be,

$$\cos(ka) = \cos(\alpha a) + \frac{P\sin(\alpha a)}{\alpha a};\; \text{where } P = \frac{mV_0 ba}{\hbar^2} \quad (1.22)$$

We obtain the following for energy values inside the well (E<0) as,

$$\cos(ka) = \cos(\beta a)\cosh[\alpha(a-b)] - \frac{\beta^2 - \alpha^2}{2\alpha\beta}\sin(\beta b)\ \sinh[\alpha\ (a-b)]$$

with $\alpha^2 = \frac{2m|E|}{\hbar^2}$ and $\beta^2 = \frac{2m(V_0 - |E|)}{\hbar^2}$

Using the same estimates as above *($b \to 0$; $V_0 \to \infty$; $V_0 b$ = constant)*,

we arrive at,

$$\cos(ka) = \cosh(\alpha a) + \frac{P\sinh(\alpha a)}{\alpha a} \tag{1.23}$$

Utilising the same P formula as in the prior case,

$$P = \frac{mV_0 ba}{\hbar^2} \tag{1.24}$$

### 1.2.2 Band Gaps in the Kronig-Penney Model

The crystal momentum k and the energy E are the only variables in the above section that are not governed by the physical system's properties.

Calculating the right-hand side of the equation 1.23 by choosing a value for E, and then one can calculate k by taking arccos (inverse of cosine function) of both sides. This expression gives leads to the dispersion relation. The Kronig-Penney model's dispersion relationship, with P = 1.5 is shown in Fig. 1.6. The value of the expression in the dispersion relation, with P = 1.5, to which cos(ka) is equal.

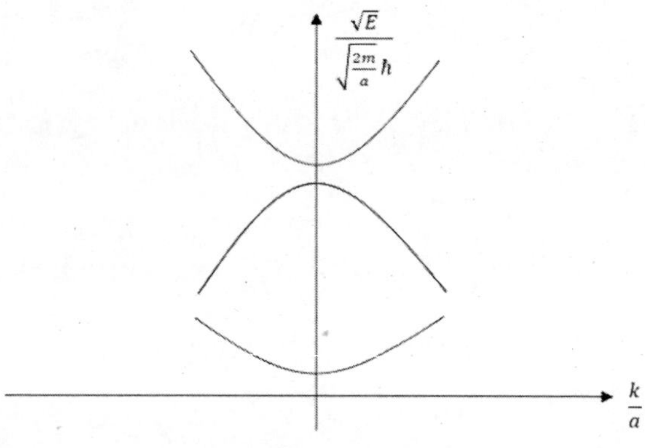


**Fig.1.6:** The dispersion relation of the Kronig-Penney model, with P=1.5

The final phrase in this equation 1.23, can occasionally have a right-hand side higher than 1 or less than -1; in such cases, the equation cannot be true for any value of k. Since α is directly proportional to indicates that, the Schrodinger equation does not have eigen functions for certain values of E. The band gap is made up of these values. Thus, one of the most basic periodic potentials, to display a band gap is the Kronig-Penney model.

**Example 1.2:** The wave function $\psi(x) = C\sin(kx)$ for (0, L) where L is the width of the potential well which satisfies the Schrodinger equation for an infinite 1-D potential. What is the value of C. (Hint: The overall chance of detecting the particle at any x should be one, hence this information may be used to derive the normalisation constant C)

**Given:** Wave function $\psi(x) = C\sin(kx)$

**Solution:**

We know that, $\int_{-\infty}^{\infty} |\psi(x)|^2\, dx = 1$

Substitute the wave function we get,

$$\int_0^L C^2\sin^2(Kx) = 1 \rightarrow C^2 \int_0^L \frac{1-\cos 2(kx)}{2}.\ dx = C^2\left[x - \frac{\sin 2(kx)}{2kx}\right] = 1$$

$$\frac{C^2}{2}.\ L = 1$$

$$C = \sqrt{2/L}$$

## 1.3 ENERGY BANDS IN INTRINSIC AND EXTRINSIC SILICON

Energy levels of electrons in each orbit merge each other to form an energy band. Electrons in a similar orbit demonstrate various levels of energy. The energy levels of valence electrons merge with each other to form a valence band. According to Bohr's hypothesis, each atom's shell holds a distinct quantity of energy at various levels. When a valence electron absorbs energy, it becomes a free electron. Electron interaction between the outermost and innermost shells are explained by the energy band theory. The energy levels of all the free electrons merge each other to form a *conduction band*. The *forbidden band* is the name of the energy band that lies in between the valence band and the conduction band. The valence band electrons are still under the force of attraction of the nucleus. The electrons in the conduction band are free from the force of attraction of the nucleus.

In general, no electrons exist in the forbidden band. The energy difference between the energy of the conduction band and the valence band is termed as the *forbidden energy gap* or *band gap* (EG).

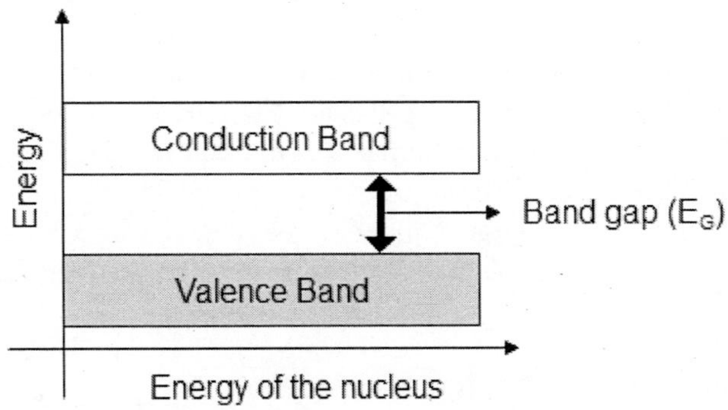


**Fig.1.7:** Energy Band Diagram

This represents the amount of energy required to be given to electrons of the valence band, to transfer them to the conduction band. The graphical representation of the energy band diagram is shown in Fig. 1.7. The energy gap is measured in the unit of "*electron-volt*" (eV).

## 1.3.1 Classification Based on Energy Band Diagram

Based on the band gap of the material, the materials can be categorized as semiconductors, insulators, and conductors.

- In conductors, an abundance of free electrons exists at normal room temperature. So, no energy gap is present between the conduction and valence band. Both conduction and valence bands are overlap. A conductor is represented by an energy band diagram in Fig. 1.8 (a). Examples of conduction materials are copper, aluminum, silver, etc.,
- In an insulator, the band gap is very high about 9eV. Hence at very high voltage or temperature, the electrons cannot move toward the conduction band. So, these materials are known as insulators, as they had zero conductivity. The energy-band representation of the insulator is illustrated in Fig. 1.8 (b). Examples are wood, paper, mica, etc.,

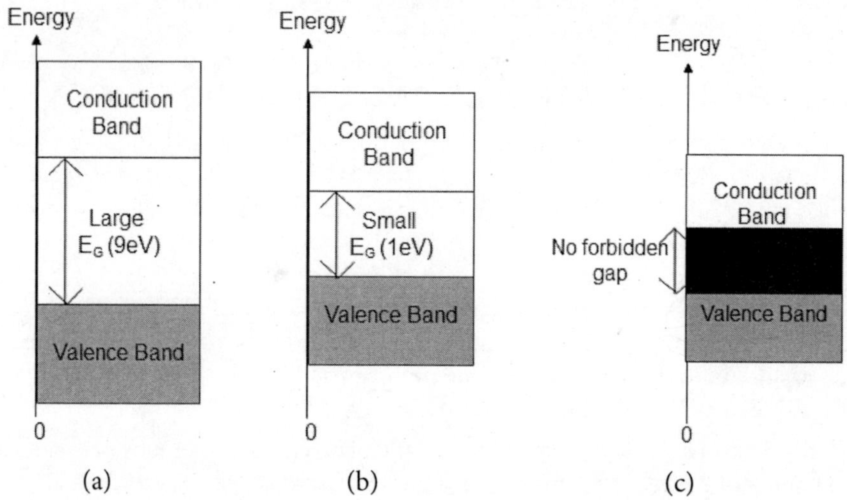


**Fig.1.8:** Energy Band Diagram (a) Conductor, (b) Insulator, (c) Semiconductor

- In semiconductors, the conduction band is said to be in an empty state, and the valence band contains electrons. The energy gap for silicon is about 1.12eV and for Germanium is 0.78eV. The electrons transition from the valence band to the conduction band occurs at room temperature. Hence the material starts to conducting partially. As temperature increases, the band gap starts decreases and a large number of free electrons are available. Example materials are silicon and Germanium. The energy band diagram of the semiconductor is shown in Fig. 1.8. (c).

## 1.3.2 Structure of Semiconductor Material

The atomic structure of silicon and germanium are shown in Fig. 1.9. The atomic number of silicon is 14 and for Germanium is 32. Both structures consist of 4 valence electrons in the outermost shell.

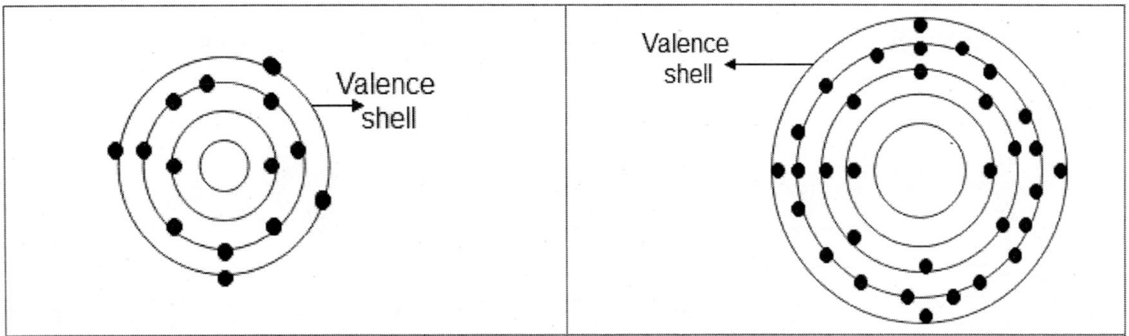


(a) (b)

**Fig.1.9:** Atomic Structure of (a) Silicon, (b) Germanium

## 1.3.3 Intrinsic Semiconductor

A sample of a semiconductor in its *purest* form is called an *intrinsic semiconductor.* The conductivity of such type of a semiconductor is very poor and practically cannot be used for manufacturing of semiconductor devices. In the outermost shell of the intrinsic semiconductor, there are 4 electrons. Each atom shares its valence electrons with its neighbour atoms and forms covalent bonds. Hence, the outermost shells of all the atoms are completely filled with 8 electrons. So, there is a no free electron, and the conductivity of intrinsic semiconductor is very worst. At room temperature, intrinsic material behaves as an insulator. Therefore, intrinsic semiconductors are not suitable for electronic device manufacturing.

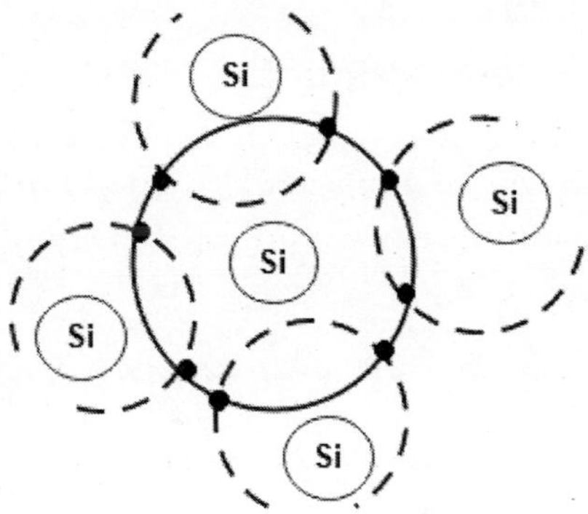


**Fig.1.10:** Crystalline Structure of Silicon

### *1.3.3.1 Energy Bands in Intrinsic Semiconductor*

In the energy band diagram, the probability of occupancy of an energy is indicated by a level called *fermi level* denoted as . In intrinsic semiconductor,

No. of electrons = No. of holes

In the band diagram,

- lowest conduction band energy level
- highest valence band energy level

Probability of finding electron in conduction band is equal to probability of finding holes in valence band. In intrinsic semiconductor, the fermi energy level is in the middle of the forbidden gap. Band diagram of intrinsic structure is shown in the Fig. 1.11. Fermi level $E_F$ is given by, $E_F = \frac{E_C + E_V}{2}$

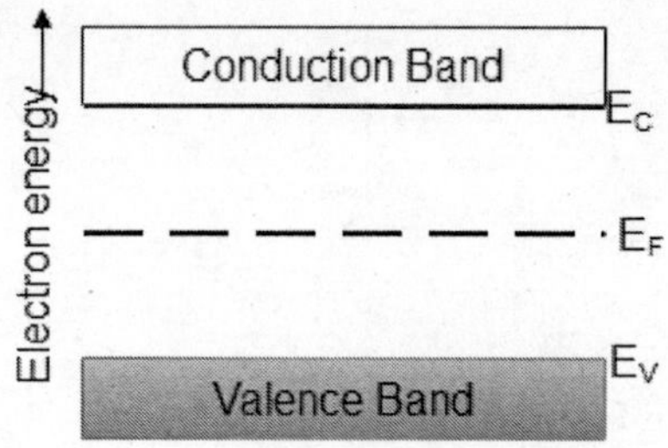


**Fig.1.11:** Energy Band Diagram of Intrinsic Semiconductor

## 1.3.4 Extrinsic Semiconductor

Extrinsic semiconductors are defined as semiconductors wherein certain dopant or impurity atoms have been added in a predetermined ratio. So that, the intrinsic carrier concentration is not maintained while the thermal equilibrium electron and hole concentrations are altered. One type of carrier either p-type or n-type, will predominate in an extrinsic semiconductor. "*Doping*" is the process of introducing impurities to the intrinsic semiconductor. There are two types of impurities used to obtain two different types of extrinsic semiconductors called "*p-type*" and "*n-type*".

### *1.3.4.1 Formation of n-type Semiconductor*

A *pentavalent* impurity has five valence electrons. When this is combined, each atom provides a free electron, and this type of doping is known as *donor doping*. Arsenic, bismuth, and phosphorous are pentavalent impurity examples. This type of doping creates an n-type extrinsic semiconductor.

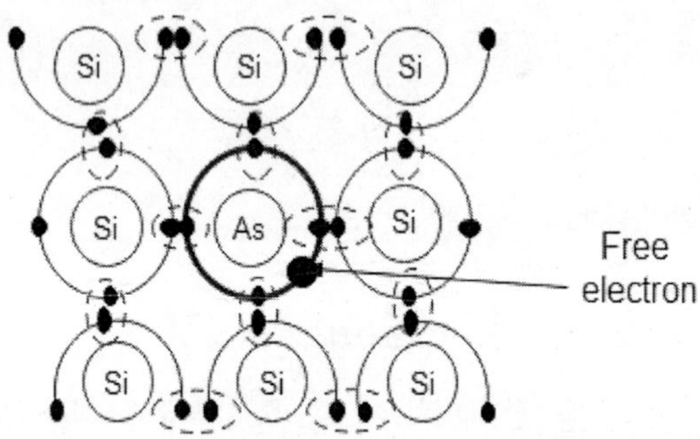


**Fig.1.12:** n-type Semiconductor Formation

Consider an arsenic atom as a doping element, the silicon atom is positioned in the crystal structure. So, that its four valence electrons may make covalent connections with four other silicon atoms that are next to it. Thus, every atom combined with Si donates one free electron shown in Fig. 1.12. The number of valance electrons is controlled by the amount of impurity added.

### 1.3.4.2 Energy Band Diagram of n-type Semiconductor

When impurities are added, allowable energy levels are introduced. In n-type material, each donor atom donates one free electron and hence donor energy level is introduced just under the edge of the conduction band denoted as $E_D$.

The conduction band and the donor energy level are extremely close together. So, all the added electrons move into the conduction band. Thus, the occupancy of energy level by an electron is very high near the conduction band. Hence, the fermi level $E_F$ shifts towards the conduction band in n-type material.

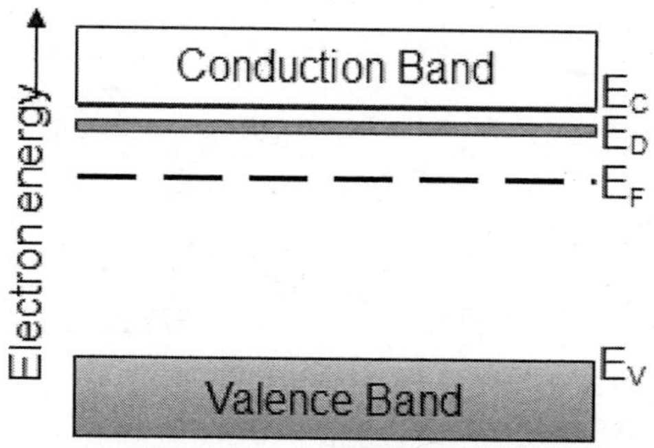


**Fig.1.13:** Energy Band Diagram of n-type Material

### 1.3.4.3 Formulation of p-type Semiconductor

Let us take a *trivalent* impurity like Boron (B) is added to Silicon (Si). When this is added, each atom accepts one free electron, and such doping is known as *"Acceptor doping"*. The way the Boron atom fits within silicon allows for the formation of covalent connections between its three valence electrons and the three neighbouring silicon atoms. There is a shortage of one electron to form a covalent bond. This creates a vacancy in the fourth covalent bond which is nothing but a *"hole"*. Thus, each Boron atom added to the Si atom creates one hole which is ready to accept an electron. The material created is referred to extrinsic p- type semiconductor.

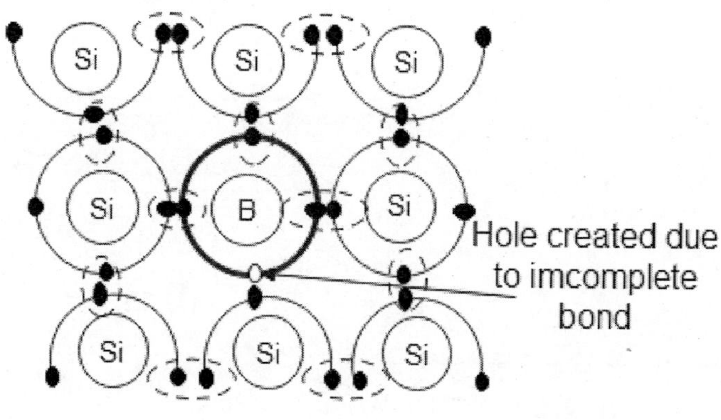


**Fig.1.14:** p-type Semiconductor Formation

### 1.3.4.4 Energy Band Diagram of p-type Semiconductor

In p-type material, each acceptor atom creates a hole in the valence band. This introduces the acceptor level of energy very near to the band of valence. It is denoted as $E_A$. The valence band develops a hole, when electrons accelerated from valance band and move to the acceptor level of energy at ambient temperature. As seen in Fig. 1.15, this causes the fermi level $E_F$, to move towards the valence band. The probability of occupying energy level is very high near the valence band.

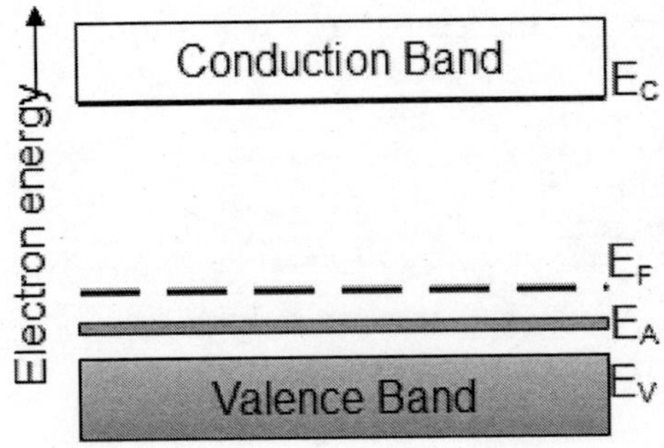


**Fig.1.15:** Energy Band Diagram of p-type Material

**Table 1.1:** Comparison between Intrinsic and Extrinsic semiconductor

| S.No | Intrinsic Semiconductor | Extrinsic Semiconductor |
|---|---|---|
| 1 | No doping is added. It is purest form of semiconductor. | Doping is added to make the semiconductor impure. |
| 2 | It has a low conductivity. | Comparatively speaking, it has a greater conductivity than an intrinsic semiconductor. |
| 3 | Conduction and valence bands have a narrow band gap. | Compared to intrinsic semiconductor, the energy gap is larger. |
| 4 | It is found in that prohibited energy gap at the centre. | According to the kind of extrinsic semiconductor, there are different degrees of fermi level existence. |
| 5 | The temperature is a factor in conduction. | Temperature and the amount of doped impurity present affect conduction. |
| 6 | Both bands contain same number of holes and electrons. | Depending on the kind of extrinsic semiconductor, electrons and holes are present in greater quantities. |
| 7 | No additional classification is offered. | It falls within the p-type and n-type semiconductor categories. |
| 8 | Some examples are Si, Ge etc. | Examples for extrinsic semiconductor are GaAs, GaP etc. |

## 1.4 Energy - Momentum Diagram (E-K Diagram)

The earlier parts explored the fundamental ideas of permissible and restricted energy bands. The Energy-Momentum (E-K) diagram illustrates a certain property of semiconductor material. It demonstrates the connection between the energy and momentum of the possible quantum mechanical states for the material's electrons. K is the momentum and E is the energy from a mathematical point of view. E-K diagrams of semiconductors are obtained by solving Schrodinger's equation. The effective mass is given by,

$$\frac{1}{m^+} = \frac{1}{\hbar^2}\frac{d_2E}{dk^2} \tag{1.25}$$

The minimal conduction band energy, which is at =0, is where the electrons in the conduction band typically settle. Similar to this, the topmost valence band energy is where holes in the valance band tend to gather. The bottom of the conduction band and top of the valence band lies at the same value of , as illustrated in Fig. 1.16 (a). This is known as *direct bandgap semiconductor.* one that possesses this characteristic, there is no change in crystal momentum while transitioning between the two allowed bands. The optical characteristics of the materials are greatly impacted by this direct nature. GaAs is an illustration of a direct bandgap material. These kinds of components are used in optical devices like semiconductor lasers.

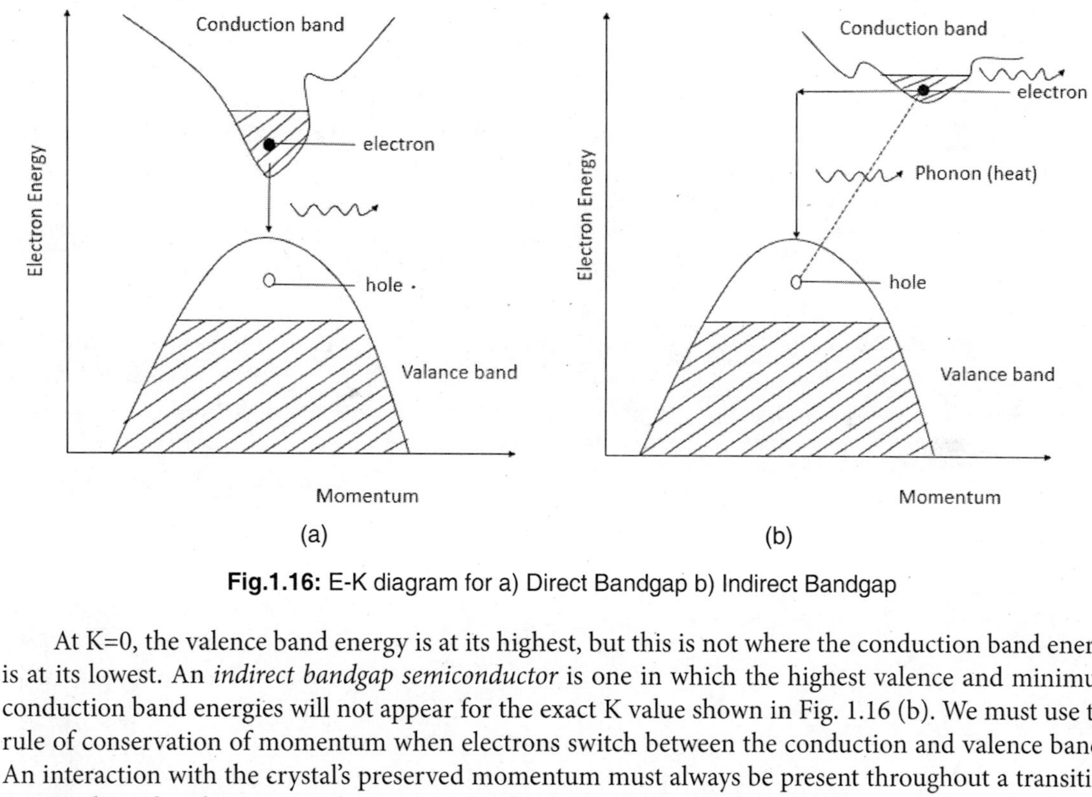


**Fig.1.16:** E-K diagram for a) Direct Bandgap b) Indirect Bandgap

At K=0, the valence band energy is at its highest, but this is not where the conduction band energy is at its lowest. An *indirect bandgap semiconductor* is one in which the highest valence and minimum conduction band energies will not appear for the exact K value shown in Fig. 1.16 (b). We must use the rule of conservation of momentum when electrons switch between the conduction and valence bands. An interaction with the crystal's preserved momentum must always be present throughout a transition in an indirect bandgap material. An example of an indirect bandgap is Silicon.

The electron effective mass is connected to versus curvature diagram. For materials with a direct bandgap, the curvature of the conduction band and conduction band minimum are larger than for materials with an indirect bandgap. Therefore, a direct bandgap material's effective mass of an electron in the conduction band will be lower than that of an indirect bandgap material.

### 1.4.1 Significance of E-K Diagram

- No theoretical study, experimentation, and technological application can take place without an E-K diagram.
- This diagram indicates the bandgap $E_g$ which is the difference in energy between the top of the valence band and bottom of the conduction band.
- It explains electron (hole) effective mass and mobility.
- It indicates how the electron states are equally spaced in K-Space.
- The E-K diagram clearly depicts the direct vs indirect band gap.

## 1.5 CARRIER TRANSPORT

These are the mechanisms by which the electron and hole concentrations could vary in the application of the electric field. Based on the variation or the count of electron and hole concentration, we can establish what amount of current would be in semiconductors. This carrier transport mechanism is easily defined using the continuity equation. Diffusion and drift are two charge carrying mechanism.

### 1.5.1 Drift Current

*Drift current* is used to describe the flow of charge carriers caused by an electric field or voltage that has been applied. We can classify the charge carriers into two different types: electrons and holes. Free electrons migrate towards a battery's positive terminal for a voltage supplied to a semiconductor, whereas holes move in the opposite direction. Holes are positively charged particles and negatively charged ones are electrons. We are aware that opposite charges attract one another whereas similar charges repel one another. As a result, holes (positively charged particles) are drawn to a battery's negative terminal, while electrons (negatively charged particles) are drawn to its positive terminal as shown in Fig. 1.17.

The electrons in a semiconductor often attempt to travel in a straight line in the direction of the battery's positive terminal. But they alter the direction of flow as a result of constant interaction with the atoms. The electron always returns in a random path after striking an atom. While the collision, random movement of electrons are not stopped by the applied voltage, it does cause the electrons to drift toward the direction of the positive terminal.

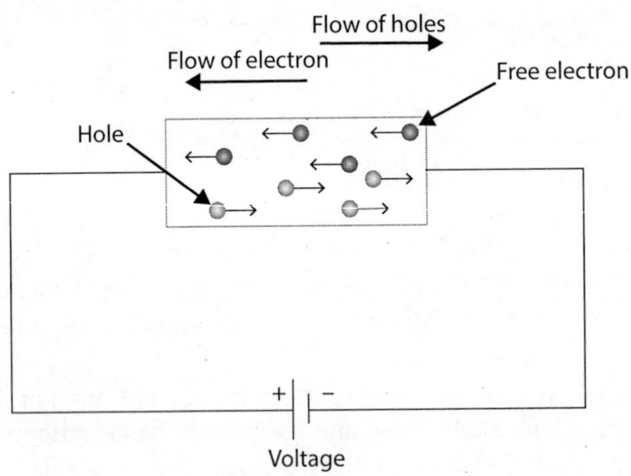


**Fig.1.17:** Drift mechanism of charge carriers

*Drift velocity* is an average speed that an electron or hole travelled at an electric field / applied voltage.

The electron's *drift velocity,* $V_n$ is given by,

$$V_n = \mu_n E \tag{1.26}$$

The *drift velocity of holes,* $V_p$ is given by,

$$V_p = \mu_p E \tag{1.27}$$

where, $\mu_n$ is electron mobility, $\mu_p$ is hole mobility and E is the field applied on the material.

### 1.5.1.1 Drift Current Density

Drift current density refers to the movement of electric charge carriers, such as electrons or holes, in a semiconductor material due to an electric field. It is the current density caused by the drift of these charge carriers and is expressed in units of amperes per square meter ($A/m^2$). The drift current density is proportional to the carrier concentration and the drift velocity, which is the average velocity of the charge carriers in response to the electric field. The total drift current density is denoted as $J_d$.

The free electron drift current density, $J_n$ is given by,

$$J_{ndrift} = en\mu_n E \tag{1.28}$$

The hole-induced drift current density, $J_p$ is defined as,

$$J_{ndrift} = en\mu_p E \tag{1.29}$$

where,

$e = 1.6 \times 10^{-19}$ coulombs (C)

n = electron concentration

p = hole concentration

Overall drift current density,

$$J_d = J_{ndrift} + J_{pdrift} \tag{1.30}$$

$$J_d = en\mu_n E + ep\mu_p E \tag{1.31}$$

$$J_d = e(n\mu_n + p\mu_p)E \tag{1.32}$$

Drift current density is directly proportional to carrier drift velocity, carrier concentration and carrier charge.

$$J_n = qnV_{dn} \tag{1.33}$$

$$J_p = qpV_{dp} \tag{1.34}$$

## 1.5.2 Diffusion Current

Consider the n-type semiconductor in Fig. 1.18 below, which has non-uniform doping. Due to the concentration gradient, the charge carriers have a tendency to move from of higher concentration to lower concentration. As a result, the semiconductor material acquires equal electron concentration. So, the current starts to the flow from one side to the other of the semiconductor. This current is known as *Diffusion Current* shown in Fig. 1.19.

**Fig.1.18:** Non-uniform Doping Concentration

The diffusion process operates similarly in p-type semiconductors. In semiconductor devices, drift and diffusion current both exist. Without the application of an external electric field or voltage, diffusion current occurs. In a conductor, there is no diffusion current. Diffusion current flows in a direction that is similar to or opposite from drift current.

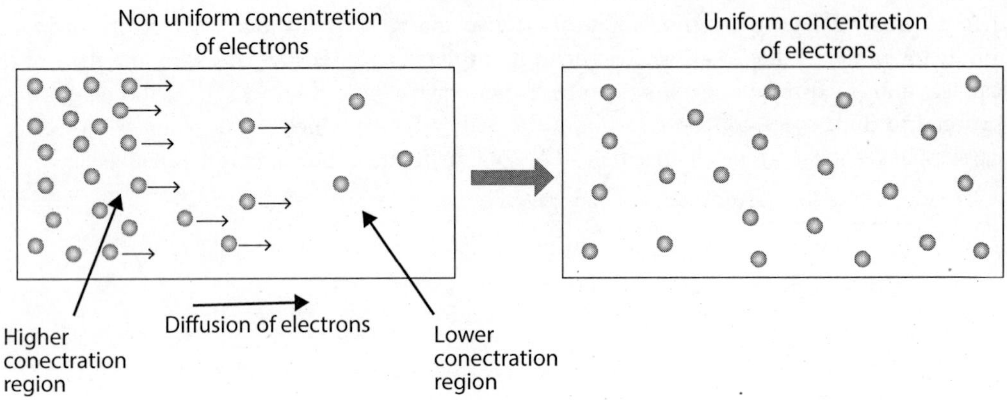


**Fig.1.19:** Diffusion of carriers from higher concentration to lower concentration

### *1.5.2.1 Diffusion Current Density*

The difference in the concentration of electrons or holes in a particular location is known as a concentration gradient. The concentration gradient directly affects the diffusion current density. Diffusion current density is high in the presence of a strong concentration gradient and vice versa. The current density of diffusion is low if the doping concentration is small.

For a N-type semiconductors, the concentration gradient is given by,

$$J_n \alpha \frac{dn}{dx} \tag{1.35}$$

For p-type semiconductors, the concentration gradient is given by,

$$Jp\alpha \frac{dp}{dx} \tag{1.36}$$

Charge and carrier flow combine to create diffusion current density.

$$J_{ndiff} = qD_n \frac{dn}{dx} \tag{1.37}$$

$$J_{pdiff} = qD_p \frac{dp}{dx} \tag{1.38}$$

### 1.5.3 Total Current Density

Total Current density (J) is given as,

$$J = J_n + J_p$$

$$J_n = J_{ndrift} + J_{ndiff} = en\mu_n E + qD_n \frac{dn}{dx} \quad (1.39)$$

$$J_p = J_{pdrift} + J_{pdiff} = en\mu_p E + qD_p \frac{dp}{dx} \quad (1.40)$$

**Table 1.2:** Comparison between Drift Current and Diffusion Current

| S.No | Drift Current | Diffusion Current |
|---|---|---|
| 1 | Drift current is a term used to describe the movement of charge carriers caused by an applied electric field. | Movement of charge carrier due to concentration gradient. |
| 2 | The process of drift current requires electrical energy. | Diffusion current may operate with a small quantity of external energy. |
| 3 | Ohm's Law is abided by the drift current. | The present complies with Fick's Law. The diffusion densities for charge carriers are symbolically opposite one another. |
| 4 | In a semiconductor, the direction of the charge carriers is the opposite of one another. | The concentration of the carrier slope can determine this current's direction. |
| 5 | The polarity of the applied electric field has a significant impact on the direction of this current. | The charge within the carrier concentrations mostly determines this current's direction. |

**Example 1.3:** Consider area $10^{-6}$ $m^2$ of copper wire with 2 Ampere current and free electrons 8 × $10^{28}$ $m^3$. Find current density (J) & average drift velocity ($v_d$).

**Given:** Area (A) = $10^{-6}$ $m^2$, current (I) = 2A

**Solution:**

$$\text{Current density } J = \frac{1}{A}$$

$$= \frac{2A}{10^{-6} m^2} = 2 \times 10^6 \ Am^{-2}$$

We know that, $I=nAev_d$

Then the average drift velocity can be written as,

$$v_d = \frac{1}{nAe}$$

$$= 2/(8 \times 10^{28} \times 10^{-6} \times 1.6 \times 10^{-19})$$

$$v_d = 1.56\,0 \times 10^{-4} \ m/s$$

## 1.6 MOBILITY AND RESISTIVITY

Mobility determines how fast a carrier is movable inside the semiconductor material with an applied field. When there is no applied field on the semiconductor, the carriers start to move in a random manner, and there is no net flow of electrons. So, there is no current flow. When we apply a small electric field (E), a random thermal collision takes place and the electron starts to move in an opposite direction to the applied field with the velocity as shown in Fig. 1.20. This velocity is known as drift velocity.

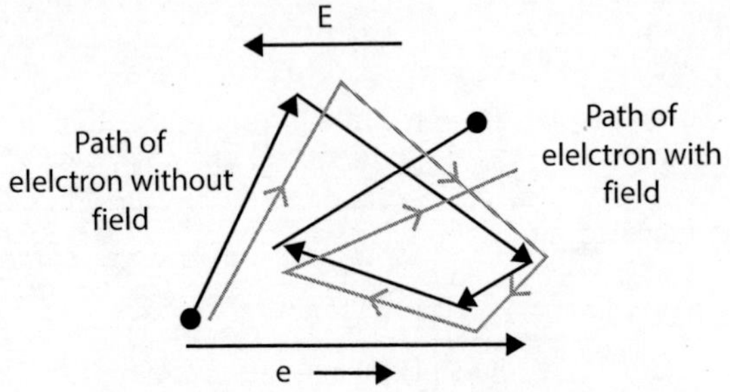


**Fig.1.20:** Thermal motion of electron

The drift velocity of electron is determined by,

$$v_d = \frac{a\tau}{m^*} \tag{1.41}$$

where is the acceleration of carrier and it is given by, , is the mean collision time and is the effective mass of the carrier. Substituting in equation 1.41.

$$v_d = \frac{(qE)\tau}{m^*} \tag{1.42}$$

$$v_d = \frac{q\tau}{m^*}.E \tag{1.43}$$

The equation 1.43 states that, mobility relates the average carrier drift velocity to the electric field. It is expressed as,

$$\frac{v_d}{E} = \mu \tag{1.44}$$

Compare equations 1.43 and 1.44, we get the mobility expression as,

$$\mu = q\tau/m^* \tag{1.45}$$

Carrier mobility is one of the most important parameters of any semiconductor material, determining the sustainability for application in a large variety of electronic device.

### 1.6.1 Effects of Mobility

Two basic scattering mechanism influences the mobility of charge carriers.

1. Scattering due to lattice mismatch
2. Scattering due to impurities

When the vibration of the lattice scatters the crystal, lattice scattering has taken place and crystal flaws like ionized impurities. It can cause impurity scattering. Each scattering effect is temperature dependent. For low temperature, impurity scattering dominates, whereas for high temperature lattice scattering dominates as shown in Fig. 1.21. Hole mobility is lesser than the electron. Because the effective mass of the electron is low compared to the hole.

Effective mobility,

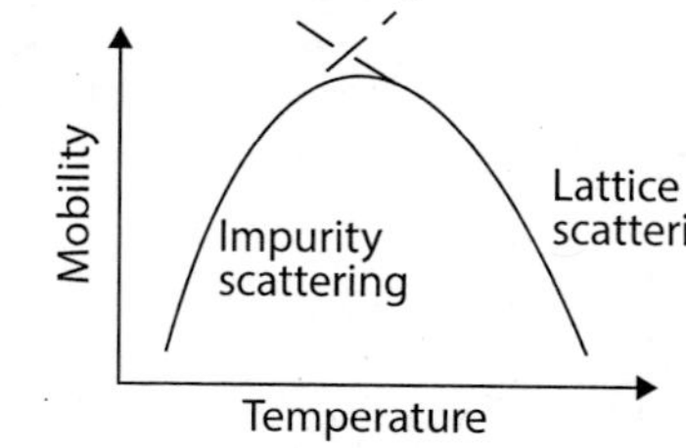


Effective mobility,

$$\frac{1}{\mu_{eff}} = \frac{1}{\mu_{lettice}} + \frac{1}{\mu_{impurity}} \quad (1.46)$$

**Fig.1.21:** Effect of mobility due to scattering

## 1.6.2 Resistivity

Resistivity is the amount of resistance, a given cross sectional area of the material will experience per unit and it is denoted by . It is depicted in Fig. 1.22. Let us consider the conductor with length (L) and area (A) having R as a resistance, the resistivity is defined as,

$$\rho = \frac{RA}{L} \quad (1.47)$$

From ohm's law,

$$J = \sigma . E \quad (1.48)$$

$$J = qn\mu\_n.E \quad (1.49)$$

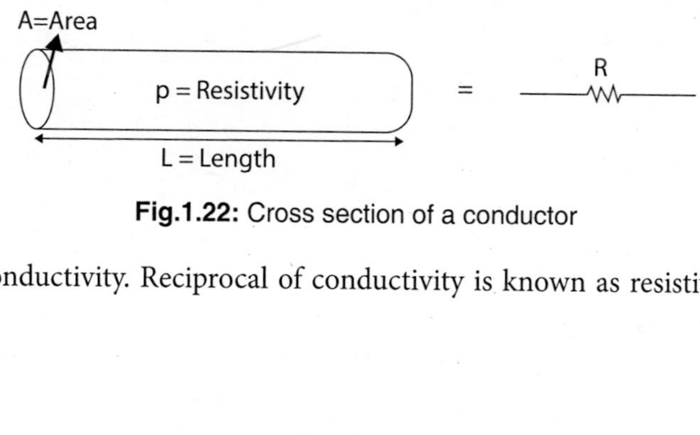


**Fig.1.22:** Cross section of a conductor

where σ is the conductivity. Reciprocal of conductivity is known as resistivity. Then resistivity is defined as,

$$\rho = \frac{1}{\sigma} = \frac{1}{qn\mu_n} \quad (1.50)$$

If the area of the conductor is large, then the resistivity of the conductor is high. The conductor's resistance will also be high if the conductor's length is long.

**Example 1.4:** Find the drift speed of copper wire which has area $1.0 \times 10^{-7}$ $m^2$ with 1.5 Ampere. (Hint: electron density $9\times10^{28}$ $m^{-3}$).

**Given:** Cross sectional area, $A = 1.0 \times 10^{-7}$ $m^2$, current = 1.5 A.

**Solution:**

We know that $I = nAev_d$

$$v_d = \frac{1}{nAe}$$

$$= \frac{1.5}{9\times10^{28}\times1.0\times10^{-7}\times1.6\times10^{-19}}$$

$$v_d = 1.042\times10^{-3}\ m/s$$

## 1.7 SHEET RESISTANCE AND DESIGN OF RESISTORS

Sheet resistance is a common electrical property used to describe thin films of conducting and semiconducting materials (also known as surface resistance or surface resistivity). A thin square of the material's lateral resistance, or the resistance between the opposing sides, is measured.

Fig. 1.23 shows the square of a material with resistivity , thickness , and cross-sectional area. This sheet resistance of a material can be written as,

$$R_s = \frac{pL}{L\times t} = \frac{p}{t}\ (\text{ihms / square}) \tag{1.51}$$

This quantity is independent of the square's size and is mostly determined by the material's diffusion characteristic. The sheet resistance , as well as the surface dimensions and, can be used to calculate the resistance of these resistors.

Now,

$$R = \rho\frac{L}{W\times t} \tag{1.52}$$

or

$$R = R_s\ \frac{L}{W}$$

The aspect ratio ($\frac{L}{W}$) is the effective number of squares contained in the resistor. Due to the medium resistivity (200Ω/square), p-type base region, a base resistor in the range of 20Ω to 300Ω can be easily fabricated. However, the emitter diffusion sheet resistance is merely on the order of 5Ω/sq. As a result, emitter resistors are typical, in the 10Ω to 1Ω range.

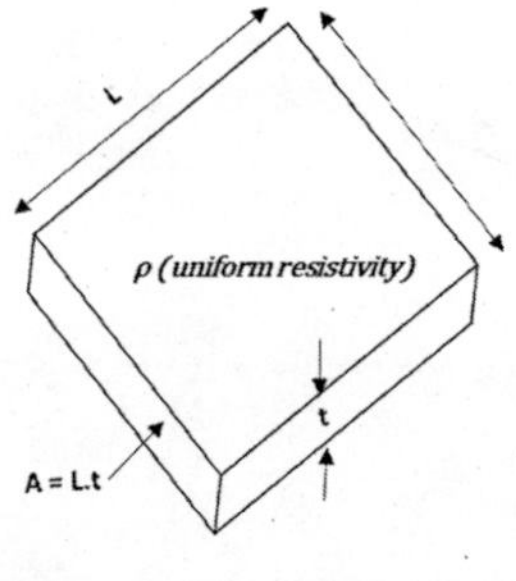


**Fig.1.23:** Cross sectional view of sheet

**Example 1.5:** The sheet resistance of p-type diffusion is 200Ω/sq. Then design an 8KΩ diffused resistor. Find the aspect ratio.

**Given:** Sheet Resistance $R_S = 200\Omega/sq$, Resistor R = 8KΩ.

**Solution:**

$$\frac{L}{W} = \frac{R}{R_s} = \frac{8 \times 10^3}{200} = \frac{40}{1}$$

Hence, a pattern of 40 miles long by 1mil wide could be used to fabricate a 8 resistor is shown in Fig. 1.24.

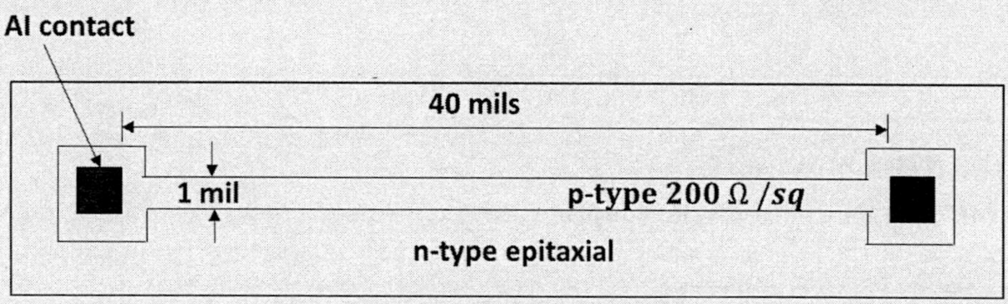


**Fig.1.24:** Design of 8Ω diffusion resistor

**Example 1.6:** A polysilicon resistor having width (W) of 0.8 μm and length (L) of 20 μm. Find resistivity ($\rho_s$ in Ω /cm), and resistance R.

**Given:** Width (W)=0.8 m, Length (*L*)= 20m.

**Solution:**

Assume that for polysilicon is $9 \times 10^{-4}\Omega$.cm and thickness of polysilicon is 3000 Å. Ignore any contact resistance.

First calculate $\rho_s$,

$$\rho_s = \frac{\rho}{T} \frac{9 \times 10^{-4}\ \Omega\text{cm}}{3 \times 10^{-5}\ \text{cm}} = 30\ \Omega\text{cm}$$

The number of squares of resistance, N is,

$$N = \frac{L}{W} = \frac{20\ \mu\text{m}}{0.8\mu\text{m}} = 25$$

Total resistance is calculated as,

$$R = \rho_s \times N = 30 \times 25 = 750\ \Omega$$

*Mils- A thousands of inches

## UNIT SUMMERY

- Planck's coefficient: $h = 6.6 \times 10^{-34}$ J.s
- Speed of light in vacuum: $c = 3 \times 10^8$ m/s

- Rest mass of electron: $m_o = 9.1 \times 10^{-19}$ kg
- Charge of an electron: = $1.6 \times 10^{-19}$ C
- De Broglie wavelength is related to the momentum 'p' and is given by,

$$\lambda = \frac{h}{p}$$

- The energy of a free electron is given by,

$$E = \frac{p^2}{2m_o}$$

- The energy of a photon is given by,

$$E = hv = \frac{hc}{\lambda} = pc$$

- **1D Potential Box of Width L**

  Consider an infinite potential well width 'L', where the electron has a mass 'm' and momentum 'p'. The motion of the particle is described by the Schrodinger equation, which in its time dependent form is given by,

$$K.E+P.E=T.E$$

$$-\frac{h^2}{2m}\frac{\partial^2\psi}{\partial c^2} + V(x)\,\psi(x) = E\psi(x)$$

- **Effective Mass**

  Effective mass accounts for the influence of the crystal lattice on the motion of the particle and connects a particle's motion in a crystal to an applied external force. The effective mass of an electron, , is generally calculated by approximating the band edge by a parabolic equation,

$$E - E_c = C_1 k^2$$

  The energy $E_c$ is conduction band. Since $E > E_c$, the value of the parameter $C_1$ is a positive. The effective mass, m*, is calculated as,

$$\frac{1}{\hbar^2}\frac{d^2E}{d^2k} = \frac{2C_1}{\hbar^2} = \frac{1}{m^*}$$

- **Equilibrium carrier concentration - Intrinsic Semiconductor**

  For an intrinsic semiconductor, the thermal equilibrium electron and hole concentration are given by,

$$n_0 = N_c \exp\left(-\frac{(E_c - E_F)}{kT}\right),\ p_0 = N_v \exp\left(-\frac{(E_F - E_v)}{kT}\right)$$

  The intrinsic carrier concentration is provided by,

$$n_i^2 = N_c N_v \exp\left(-\frac{E_g}{kT}\right)$$

The intrinsic fermi level with respect to the midgap position depends on the effective masses of electron and hole and is given by,

$$E_{Fi} - E_{midgap} = \frac{3kT}{4} \text{In}\left(\frac{m_p^*}{m_n^*}\right)$$

- **Equilibrium carrier concentrations - Extrinsic semiconductor**

$$n_0 = N_i \exp\left(\frac{(E_F - E_{Fi})}{kT}\right), \quad p_0 = n_i \exp\left(-\frac{(E_{Fi} - E_F)}{kT}\right)$$

Alternatively,

$$n_0 = \frac{N_d - N_a}{2} + \sqrt{\left(\frac{N_d - N_a}{2}\right)^2 + n_i^2}, \quad p_0 = \frac{N_a - Na}{2} + \sqrt{\left(\frac{Nd - N_d}{2}\right)^2 + n_i^2}$$

where $n_d$ and $n_a$ are donor and acceptor concentration respectively.
The mass action law is given by,

$$n_0 p_0 = n_i^2$$

The Fermi level is located relative to the intrinsic Fermi level in the following way:

$$E_F - E_{Fi} \equiv kT\text{In}\left(\frac{n_0}{n_i}\right), \quad E_{Fi} - E_F \equiv kT\text{In}\left(\frac{p_0}{n_i}\right)$$

- **Carrier drift**

The drift current density and drift velocity for a semiconductor in an applied electric field, E is given by,

$$J_{drift} = e(\mu_n n + \mu_n p)E = \sigma E$$

$$v_{drift} = \frac{q\tau E}{m^*}$$

- **Diffusion current**

$$J_{diff} = qD_n \frac{dn}{dx} + qD_p \frac{dp}{dx}$$

Expression for mobility of the charge carriers

$$\mu = q\tau/m^*$$

Where $\tau$ is the mean Collison time, $m^*$ - effective mass of the electron and q is the charge of the electron.

- **Effective mobility due to the impurity and scattering parameters**

$$\frac{1}{\mu_{eff}} = \frac{1}{\mu_{lattice}} + \frac{1}{\mu_{impurity}}$$

- **Resistivity of the conductor**

It is defined as the resistance of the material having specific dimensions.

$$\rho = \frac{RL}{A'} \qquad \rho = \frac{1}{\sigma} = \frac{1}{qn\mu_n}$$

where σ is the conductivity of the material, it is defined as,

$$\sigma_n = q(\mu_n n + \mu_p p)$$

But p<<n as holes is in minority hence, the conductivity of n-type material is defined as

$$\sigma_n = q\mu_n n$$

For p-type material n<<p as free electrons are in minority hence, the conductivity of p-type material is defined as,

$$\sigma_p = q\mu_p p$$

Design of resistor and sheet resistance

$$R_s = \frac{\rho L}{L \times t} = \frac{\rho}{t}$$

t is the thickness of the sheet.

## EXERCISE

## Multiple choice Question

**1.1 For the given semiconductor, how does the electron mobility change when effective mass of the electron is doubled**

a. Electron mobility is not affected
b. Electron mobility is doubled
c. Electron mobility gets halved
d. None of the above

**1.2 A net charge of neutral copper atom is**

a. 0
b. 9
c. +3
d. -1

**1.3 Assuming a copper atom loses its valence electron, the net charge of an atom net changes to**

a. -5
b. +1
c. 0
d. +4

**1.4 What type of force is experienced by the valence electron of copper atom towards the nucleus?**

a. Weak
b. Strong
c. None
d. Impossible to say

**1.5 How many valence electrons are there in an atom of silicon?**

a. 1
b. 8
c. 9
d. 4

**1.6 State the point form of ohm's law**

a. V=I*R
b. P=V*I
c. J=*E
d. P=*R

**1.7 How many numbers of proton does the nucleus of a silicon atom has?**

a. 8
b. 14
c. 11
d. 10

**1.8 An ordered pattern made of silicon atoms is called a**

a. Semiconductor
b. Crystal
c. Valence orbit
d. Covalent bond

**1.9 What creates the holes in an intrinsic semiconductor that are present at room temperature?**

a. Free electrons
b. Doping
c. Creating thermal energy
d. Creation of hole

**1.10 The energy level of an electron with regard to the nucleus ------------as it is transferred to a higher orbit level**

a. Increases
b. Remains same
c. Decreases
d. Depends on type of atom

**1.11 A free-electron and a hole combining forces is referred to as**

a. Doping
b. Life time
c. Recombination process
d. Thermal energy formation

**1.12 Dual nature of matter was predicted by**

a. Schrodinger
b. Louis De-Broglie
c. Thomson
d. Werner Heisenberg

**1.13 The interval among the creation of a hole and its closing is called**

a. Doping
b. Life time
c. Process of recombination
d. Valence bond

**1.14 Another name for a conductor's valence electron is**

a. Bound electron
b. Free electron
c. Nucleus
d. Proton

**1.15 How many different forms of flow are there in a conductor?**

a. 1
b. 0
c. 13
d. 7

**1.16 How many varieties of flows are there in a semiconductor?**

a. 2
b. 1
c. 8
d. 6

**1.17 Whenever a semiconductor is subjected to a voltage, holes start to flow**

a. Far from the -ve charge
b. In the direction of +ve charge
c. In the outside circuit
d. None of them

**1.18 When a semiconductor contains------------, the valence shell becomes saturated**

a. 0 ions
b. Contains both ions
c. 4 holes
d. 8 electrons

**1.19 The hole density of intrinsic semiconductor**

a. Is equal to the quantity of free-electrons
b. Is higher than quantity of unoccupied electrons
c. Is less than how many free-electrons there are
d. None of the above

**1.20 The temperature of absolute zero is**

a. 270°C
b. 0°C
c. 50°C
d. 25°C

**1.21 When the temperature is 0 degrees, an intrinsic semiconductor possesses**

a. Limited free electrons
b. More number of holes
c. More number of electrons
d. No holes or free electrons

**1.22 An intrinsic semiconductor has an electrical property that is**

a. Limited free electrons
b. More number of holes
c. More number of electrons
d. No holes

**1.23 What happens to an intrinsic semiconductor, electrons and holes decreases as the temperature**

a. Increases
b. Decreases
c. Remains same
d. None of the above

**1.24 Given the direction of the valence electron flow, the holes must be passing through**

a. Opposite to electron
b. Same direction as electron
c. Both direction
d. None of the above

**1.25 Hole conduction is**

a. Atom
b. Crystals
c. -ve charge
d. +ve charge

**1.26 How many numbers of valence electrons should a trivalent atom have?**

a. 0
b. 9
c. 3
d. 7

**1.27 How many valence electrons are there on an acceptor atom?**

a. 4
b. 3
c. 7
d. 10

**1.28 Which of these would you utilise to create a n-type semiconductor?**

a. Acceptor atom
b. Donor atom
c. Both (a. and (b.
d. Scandium

**1.29 How many electrons do silicon atoms in crystals have in their valence orbits?**

a. 0
b. 4
c. 12
d. 11

**1.30 Atoms that are negative ions have,**

a. Picked up a proton
b. Misplaced proton
c. Picked up an electron
d. Misplaced electron

| Answer of Multiple Choice Questions | | | | |
|---|---|---|---|---|
| 1.1 **c** | 1.2 **a** | 1.3 **b** | 1.4 **b** | 1.5 **d** |
| 1.6 **c** | 1.7 **b** | 1.8 **b** | 1.9 **c** | 1.10 **a** |
| 1.11 **c** | 1.12 **b** | 1.13 **b** | 1.14 **b** | 1.15 **a** |
| 1.16 **a** | 1.17 **d** | 1.18 **d** | 1.19 **a** | 1.20 **a** |
| 1.21 **d** | 1.22 **a** | 1.23 **a** | 1.24 **a** | 1.25 **d** |
| 1.26 **b** | 1.27 **b** | 1.28 **b** | 1.29 **b** | 1.30 **c** |

## SHORT AND LONG ANSWER TYPE QUESTIONS

### Category I

1.3 Explain, the electron mobility at which a minimum conductivity of the semiconductor occurs.

1.4 From the figure given below. Find which semiconductor material has high effective mass and which one has low effective mass.

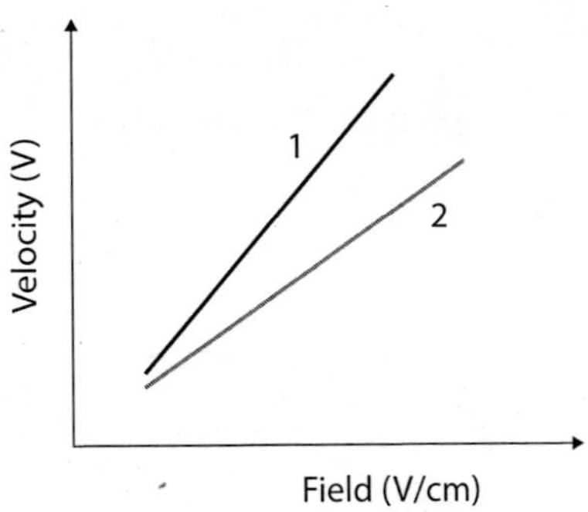


1.5 Define resistivity.

1.6 List out the scattering mechanism which affects the mobility?

1.7 What is intrinsic semiconductor?

1.8 Define an intrinsic semiconductor. Why it is not suitable for practical use?

1.9 What is doping? Explain trivalent and pentavalent dopants with suitable examples.

1.10 Define extrinsic semiconductor.

## Category II

1.1 Derive the expression for diffusion current density.

1.2 Write a note on Einstein's relation.

1.3 State and explain the law of mass action.

1.4 Explain the carrier concentrations in extrinsic semiconductor.

1.5 Obtain the n-type and p-type materials' conductivity expressions.

1.6 Derive the expression for drift current density and conductivity of a material.

1.7 Obtain the expression of conductivity of intrinsic semiconductor.

## NUMERICAL PROBLEMS

1.1 A silicon sample is 3cm long has area 0.1 $cm^2$, it has donor impurity $N_d = 2 \times 10^{15}$ $cm^{-3}$, resistance is 90Ω. Find the electron mobility?
**(Ans:104.16/$cm^{-2}$)**

1.2 Acceptor impurities are doped into a GaAs semiconductor resistor at a concentration of $N_d = 2 \times 10^{15}$ $cm^{-3}$ with area of $5 \times 10^{-5}$ $cm^{-3}$. Current of I=25mA with the bias 2V and $\mu_n$=8000$cm^2$/v – s. Calculate the length of the resistor and how much the electron drift velocity.
**(Ans: L = 0.01024cm, $v_d$= 1.56 × $10^6$ cm/s)**

1.3 Calculate the resistance of a piece of copper wire $\rho_{copper} = 1.68 \times 10^{-8}$) Ωm with a diameter of 6mm and length of 4m?
**(Ans: 7.13μΩ)**

1.4 Consider a silicon at room temperature of 270°C. Assuming $\mu_n$ = 1400 $cm^3$ /v.s and $\mu_p$ = 0.052$m^2$ /v.s, find the resistivity and conductivity if silicon is doped with a) $N_D = 6 \times 10^{16}$ /$cm^3$, b) $N_A = 5 \times 10_{16}$ /$cm^3$.
**(Ans: a) ρ = 0.074, σ = 13.44, b) ρ = 2.4KΩ, σ = 416× $10^{-6}$)**

1.5 An electron is trapped in a 1D potential box of width 20. Find the approximate energy (in ) of the emitted particle when it makes transition from the second excite state to the ground state (n=1).
**(Ans: 7.5)**

1.6 In a 1D potential well, a particle with mass m and no energy has a wave function that is independent of time, $\psi(x) = Axe^{-\frac{x^2}{L^2}}$. Determine the potential at V(0) at x=0.
**(Ans:** $-\frac{3\hbar^2}{mL^2}$)

1.7 The resistance per unit length of a piece of copper wire with circular cross section with diameter of 1mm is 2.5 × $10^6$ A/$m^2$. The current doping density is 2.1 × $10^6$ A/$m^2$. Concentration of free electrons is 8.4 × $10^{28}$ /$m^3$. Calculate current flowing, conductivity, velocity of free electrons and mobility.
**(Ans:** $1.749$ mA, $4.8 \times 10^7 (\Omega - m)^{-1}$, $1.56 \times \frac{10^{-4} m)}{s}$, $3.567 \times 10^{-3} m^2 / v - sec$ )

1.8 A negligible I is passed through a wire having length 15m and area of 6 ×$10^{-1}$ $m^2$ and its resistance to be 5Ω. What is of the material at the given temperature of the experiment?
**(Ans: 2 ×$10^{-7}$ Ωm)**

1.9 Consider the sliver wire with resistance of 2.1Ω at 27.5°C and resistance with the value of 2.7Ω at 100°C. Find the temperature coefficient of resistance of silver.

**(Ans: 0.0039°C$^{-1}$)**

1.10 A copper wire with a 0.16 cm diameter is linked in series with an aluminium wire 0.24 cm in diameter. A 10 A electric current flows through the cables. Find the aluminium wire's current density.

**(Ans: 2.2 × 10$^{6}$ Am$^{-2}$)**

## PRACTICAL

**Experiments on an Identification, Specification, Testing of R, L, C Components (Color Codes), Potentiometers, Switches (SPDT, DPDT and DIP), Bread Boards and Printed Circuit Boards (PCBs)**

### *Aim*

i. To identify the value and tolerance of resistor using color codes.

ii. To study about potentiometers, switches and bread board connection, printed circuit boards.

### *Apparatus*

- Resistors with different color code
- Potentiometers
- Bread Board
- SPDT, DPDT and DIP switches
- PCB Board

### Theory

### *Resistors*

Because resistors are too tiny to have a value inscribed on them, they are color coded. There are four or five colored bands. These range of colors are used to decipher the resistor's value. Two types resistors, that is fixed resistor and variable resistor. Fixed resistors, the value of resistor is specified and cannot be changed. Examples for fixed resistor are wire wound resistors, metal films, and carbon films.

Variable resistors that may have their value adjusted by turning the wiper. Examples for variable resistor are Semi fixed, completely variable, potentiometer.

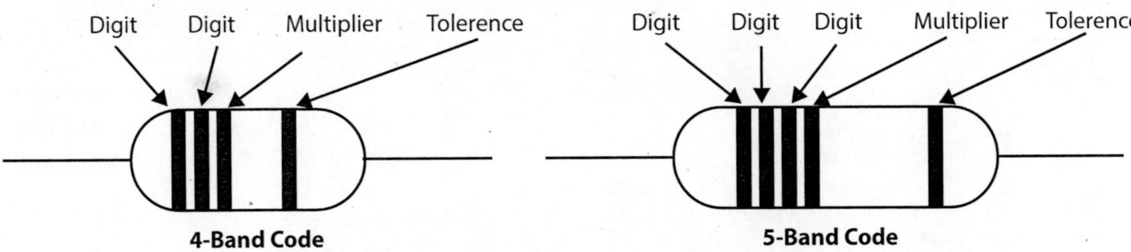


**Fig.1.25:** Representation of colour bands

## Mnemonic to Remember

"B B ROY of Great Britain had a Very Good Wife"

| Color | Digit | Multiplier | Tolerance (%) |
|---|---|---|---|
| Black | 0 | $10^0$ | |
| Brown | 1 | $10^1$ | ±1 |
| Red | 2 | $10^2$ | ±2 |
| Orange | 3 | $10^3$ | |
| Yellow | 4 | $10^4$ | |
| Green | 5 | $10^5$ | ±0.5 |
| Blue | 6 | $10^6$ | ±0.25 |
| Violet | 7 | $10^7$ | ±0.1 |
| Grey | 8 | $10^8$ | ±0.05 |
| White | 9 | $10^9$ | |
| Gold | | $10^{-1}$ | ±5 |
| Silver | | $10^{-2}$ | ±10 |
| (none) | | | ±20 |

## PROCEDURE

1. Turn the resistor such that the gold or silver band end with more bands should point left if your resistor has four different colour bands.
2. Now the left side is occupied by the first band. The initial digit is represented by this. Make a note of the numeral based on the colour. The first figure in the Fig. (ii) represents four band resistors while second figure in Fig (ii) represents the five-band resistor. The $4^{th}$ band is "5" for green. While the $5^{th}$ band is "2" for red.
3. The second digit is represented by the second band. Similar numbers to those in the first digit is represented by the colours. In this scenario, the $3^{rd}$ band is "6" for blue while the $4^{th}$ band is a "3" for orange.

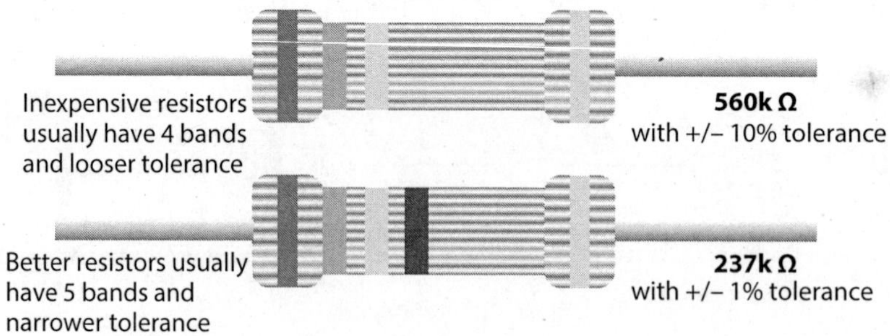


**Fig.1.26:** Resistor with 4 and 5 band colour code

4. 4. For a three-band resistor, the third band stands for the multiplier digit. The value for the multiplier band is "$10^4$". So, the value for four band resistor is 560 KΩ.
5. 4. For a resistor with five bands, the third band stands for the third digit. In this instance, the third band is "7" for violet. The second band stands for the multiplier band. the value for multiplier band is "0". The 5-band resistor's value is thus 237 ohms.
6. 5. The tolerance is shown by the last band. As a result, the 4-band resistor's value is +/-10% whereas the 5-band resistor's value is +/-1%. Another crucial characteristic to take into account is a resistor's tolerance. A resistor with a tolerance of 10% and a value of 100 ohms may actually have a fixed value between 90 and 110 ohms. A resistor with a tolerance of 10% and a resistance of 120 ohms may have a fixed resistance of 108 to 132 ohms. Therefore, there is considerable overlap between the limits of resistance at 100 and 120 ohms.

## CAPACITOR

It is a passive component, similar to a resistor. Condenser is another name for capacitor. The charge is often stored in a capacitor. The "electrical field" serves as a storage medium for the charge. In a lot of electrical and electronic circuits, capacitors are crucial.

### Capacitor in Series

Two or more capacitors are linked together in a single line when they are said to be in series. The negative plate of the capacitor after it is coupled with the positive plate of the first capacitor.

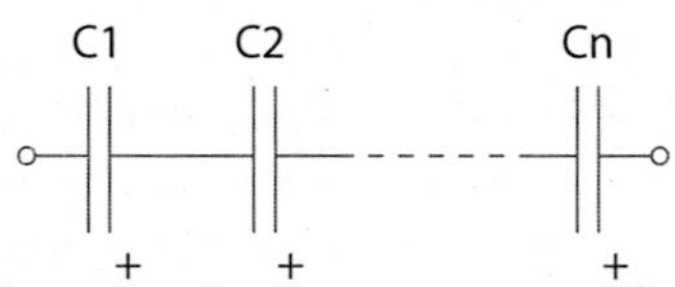


Charge and current on all of the capacitors are the same when they are linked in series. Since the charge on each plate of a series capacitor comes from the plate next to it, the same number of electrons will flow through each capacitor. So, the coulomb charge is the same. Since the movement of electrons is what causes current, current is also the same.

### Capacitor in Parallel

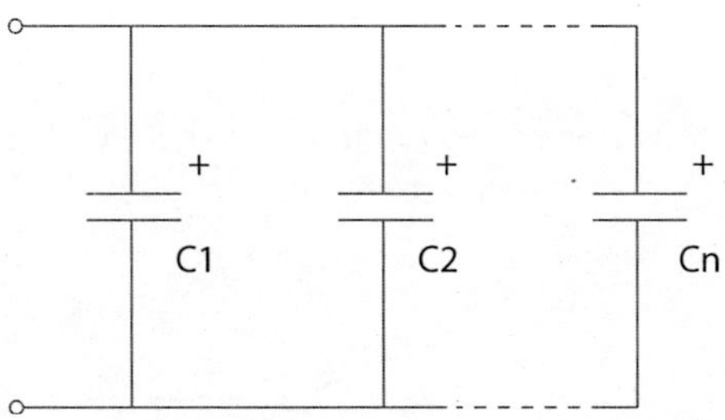


The overall capacitance increases as the capacitors are linked in parallel. Higher capacitance levels are necessary in some applications. The voltage supplied between the circuit's input and output terminals is same across all capacitors that are connected in parallel.

## Measurement using color code

Polyester capacitors have been marked with a color code for many years. Although it is no longer in use, there are undoubtedly many of them. The top three-color bands on the colors should be read like the resistor code to determine the value in pF. The fourth band (tolerance) and fifth band (voltage rating) should be ignored. Brown, black, and orange, for instance, all denote 10000pF. Due to the fact that the color bands are all the same width, there are no gaps between them. Wide red and yellow, for instance, indicate 220nF.

## Determining Capacitor Values

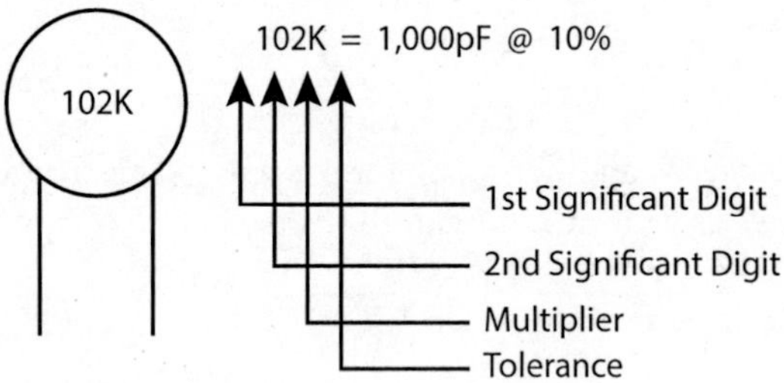


| Code | Tolerance |
|---|---|
| C | "±0.25pF |
| J | ±5% |
| K | ±10% |
| K | ±20% |
| D | ±0.5pF |
| Z | 80% / -20% |

# SWITCHES

## SPST

Understanding of poles and throws is a must for any conversation regarding switches. Poles, in essence, show how many circuits a single switch is capable of controlling. On the other hand, throws show how many contacts the switch may select from. The simplest way to comprehend this idea is via straightforward examples.

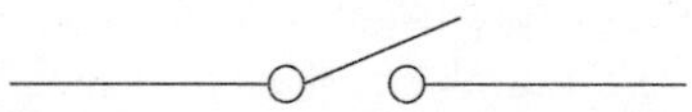

One circuit can only be controlled by a switch with a single pole and a single throw, often known as an SPST switch, and it can only open and close a single contact. That is comparable to an SPDT switch, which has a single pole but a double throw.

## SPDT

The SPDT switch can flip between two separate contacts, but there is still just one circuit that has to be managed. With an SPDT switch, it is possible to reroute the circuit rather than just opening and closing it.

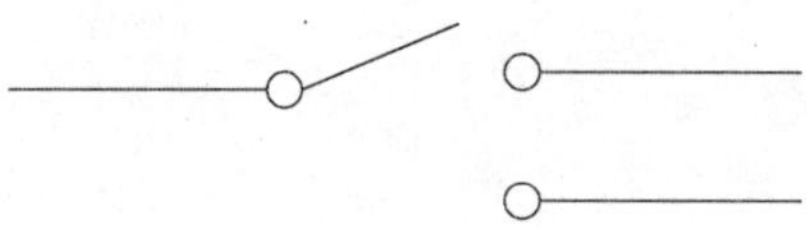

## DPDT

A double pole, double throw, or DPDT, switch controls two circuits simultaneously and alternates between two contacts on each of the two switches. The most frequent switch configurations are SPST, SPDT, DPST, and DPDT, however a switch can theoretically have as many poles and throws as it wants. In order to make things clearer, the acronyms substitute a number for the "S" or "D" when there are more than two poles or throws. For instance, the manufacturer will typically refer to a switch that has three poles and six throws as a 3P6T switch. Five tosses on a single pole might be denoted as SP5T.

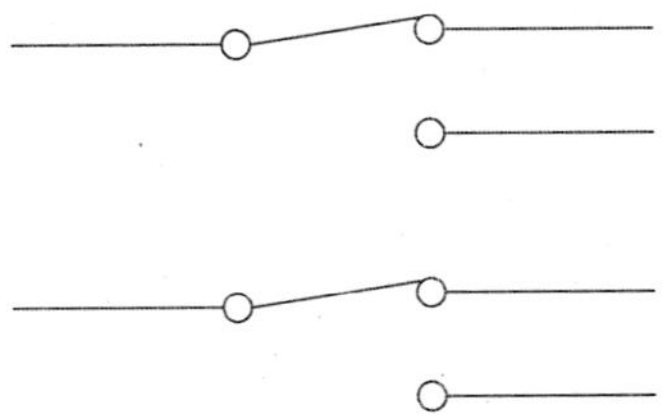

## Breadboard

A breadboard is a solderless tool used to check circuit designs and create temporary electrical prototypes. The majority of electrical elements in digital system may be linked by placing their leads on the corresponding holes, and then, when necessary, forming connections using wires.

Under the breadboard are metal strips that link the holes on top of the board. The arrangement of the metal strips is seen below. Keep in mind that while the remaining holes are connected vertically, the top and bottom rows of holes are connected horizontally and divided in half.

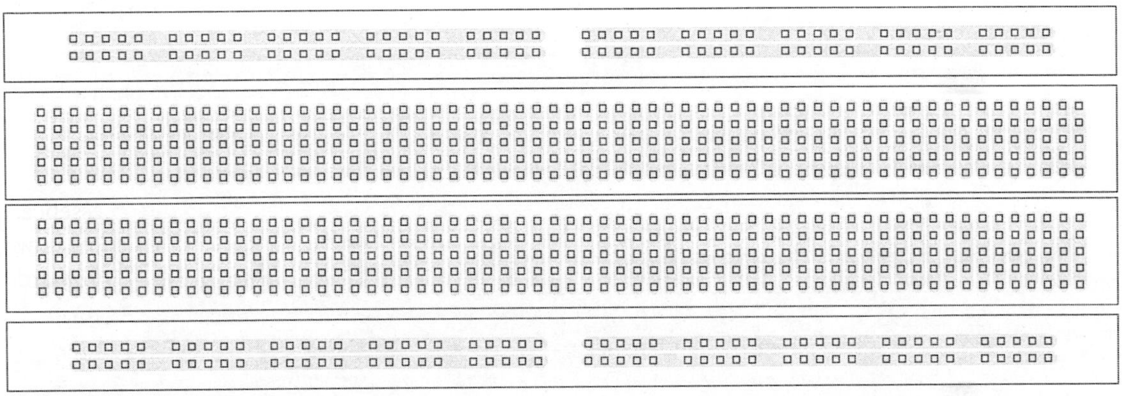

Take a note on how the holes are selected in the particular row and column are related to one another. A node is a collection of linked holes:

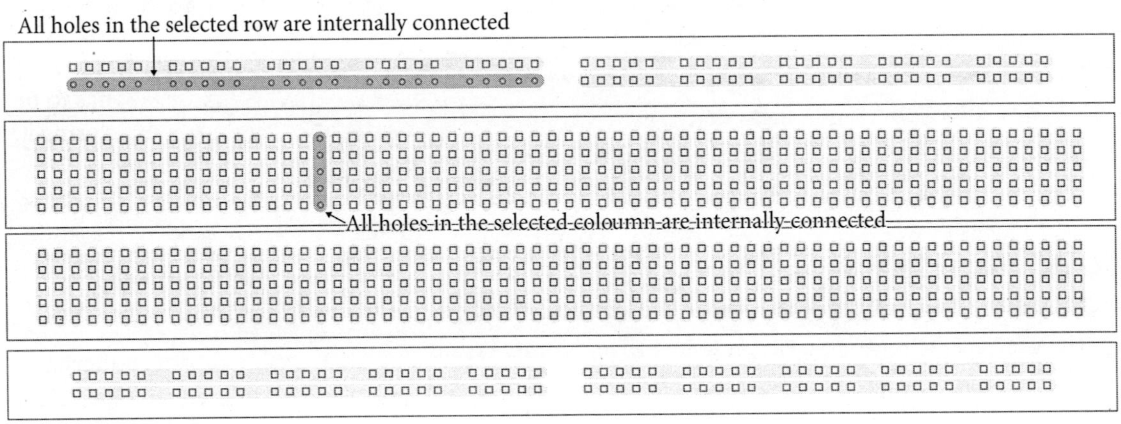

A wire connected from any hole in the row to any hole in the column is required to link the chosen row (node A) and column (node B):

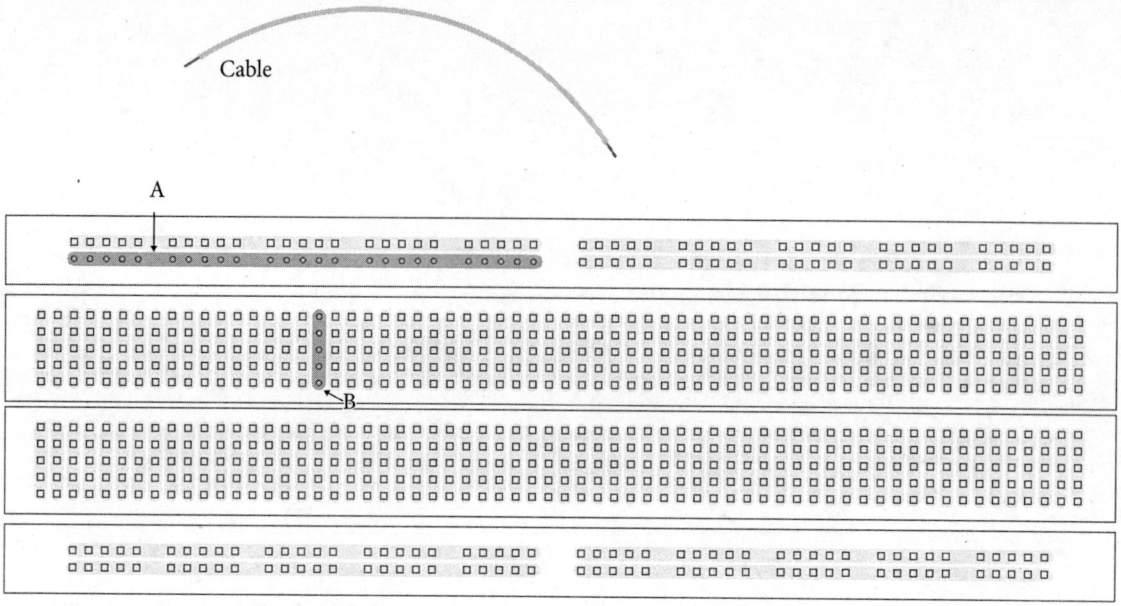


Now there is a connection between the specified row (node A) and column (node B):

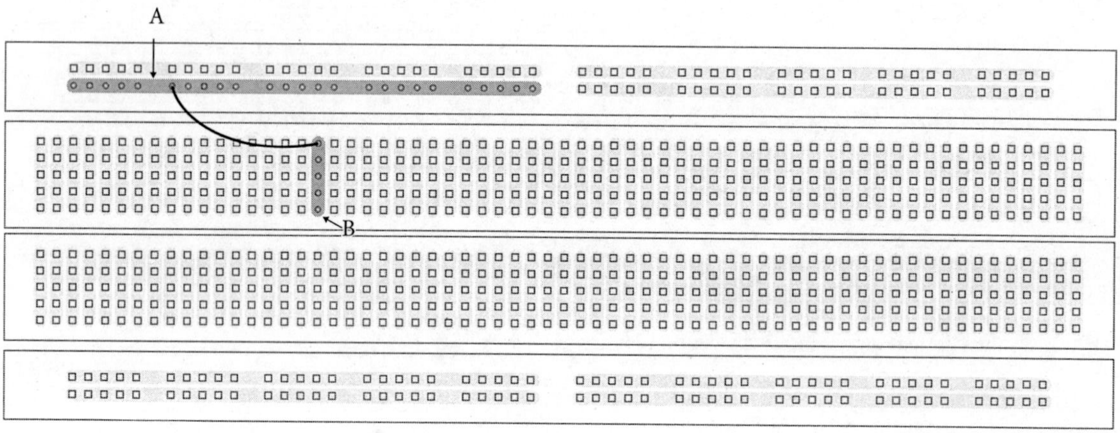


## PCB

The given schematic is converted into PCB layout.

## Procedure

1. Cut the laminate for the required size and scrub to clean it.
2. Get or draw the schematic circuit to be converted into layout.

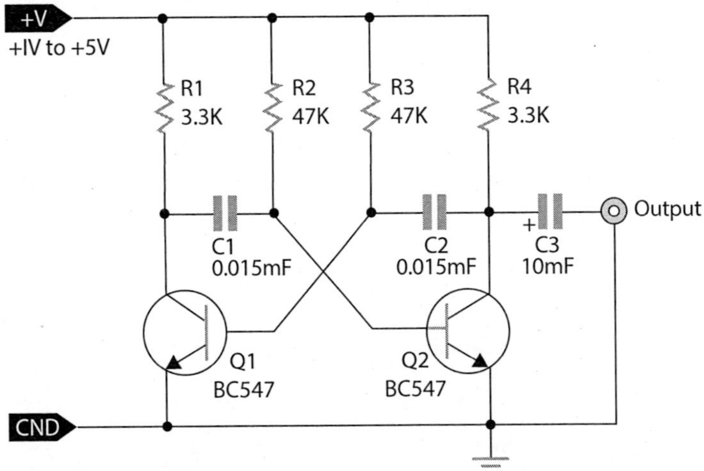


3. From the schematic circuit, identify the component package and observe the pitch reference.
4. Place (draw) the layout components in proper order as shown in the figure below.

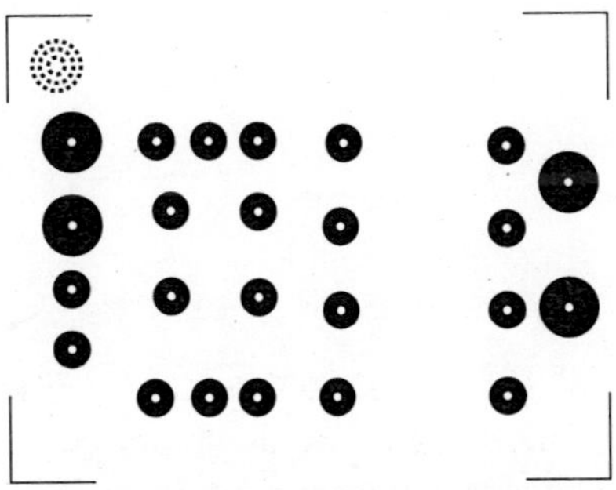

5. Connect the pad to another pad (as per the schematic reference) by a track.

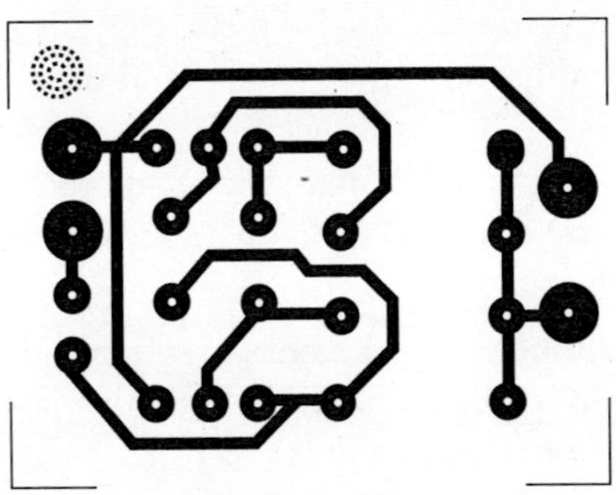

6. Mix the 25gm of ferric chloride with 100ml of water.
7. Immerse the printed laminate into the solution and agitate strongly till the unwanted copper (visible copper) is dissolved in the solution.
8. Rinse the PCB in the water and wipe the cloth.
9. Drilling the required place where the components need to be fixed.
10. Scrub and remove bur.

## KNOW MORE

"The tiniest conceivable unit of a physical entity" is what the word "quantum" refers to. Science's field of quantum physics examines the universe's tiniest objects. The theory of quantum entanglement is the interesting fact, albeit by no means the least interesting. We are aware that light moves at the greatest speed. Quantum entanglement, on the other hand, postulates that minute particles interact with one another at a pace faster than the speed of light, making it impossible for them to be described separately. This struck Albert Einstein as being quite odd, and he described it as "spooky action at a distance." The semiconductor physics is the result of this increased understanding of the atom in a lattice. Consider all the electronic gadgets you use on a daily basis: laptops, household appliances, mobile phones, headphones, fitness trackers, cars, and public transportation are all made of small pieces of technology.

### Interesting fact

It is impossible to simultaneously fully understand two properties of a system, according to a quantum principle known as the Heisenberg uncertainty principle. You can know one more exactly while knowing the other less precisely. This is true for energy and time independently as well as momentum and location. It resembles taking out a loan somewhat. You can borrow a large sum of money for a brief period of time or a smaller sum for a longer period. We then get at virtual particles. A pair of particles can briefly form if enough energy is "borrowed" from nature; they will then vanish quickly to avoid defaulting on the loan.

**Erwin Rudolf Josef Alexander Schrodinger**

Stephen Hawking thought that this process would take place near a black hole's edge, where one particle would escape (as Hawking radiation) and the other would be sucked in. As it doesn't pay back the whole amount it has borrowed, the black hole steadily disappears over time. Objects can be in two places at once. Superposition can be seen in wave-particle duality, for instance. That is, a quantum item that may be in many states simultaneously. An electron, for instance, exists simultaneously in both places. It doesn't settle into one or the other until we do an experiment to see where it is. This makes probability the central concept in quantum physics. Only after we look can we determine which state an object is most likely to be in. The wave function is a mathematical concept that contains these chances. When an observation is made, the wave function is said to "collapse," eradicating the superposition and reducing the object to just one of its many potential states.

The famous Schrodinger's cat thought experiment is resemble the concept. The fate of a cat in a locked box is connected to a quantum device. The cat is both alive and dead until we look, just as the instrument exists in both states until a measurement is taken.

### History

Since the 1830s, researchers have been studying semiconductors in lab settings. Michael Faraday began playing around with silver sulfide around 1833. He found that the material carried electricity more effectively the more it was heated. This was in contrast to how copper behaved. Copper conducts electricity less efficiently when it is heated. Other characteristics of semiconductors were identified by a number of early experimenters. The transistor was created in New Jersey's Bell Labs in 1947. As a result, integrated circuits, which power practically all modern electronic gadgets, were created.

## Application (Real Life/Industrial)

What applications do semiconductors have in modern life? It could be difficult to realise why semiconductors are utilised in so many electrical products given that they are not offered as such in stores.

For instance, semiconductors are utilised to create temperature sensors for air conditioners. Because semiconductors accurately regulate the temperature, rice cookers produce flawless results. Semiconductors are utilized to create the CPUs that power personal computers and also used in a variety of common digital consumer goods, such as mobile phones and smartphones, digital cameras, televisions, washing machines, refrigerators, and LED lights.

In addition to consumer devices, semiconductors are essential to the operation of trains, bank ATMs, the internet, communications, and other facets of the social infrastructure, including, among other things, the medical network utilised for aged care. Additionally, effective logistics systems encourage energy conservation and environmental protection. The number of semiconductor devices put on automobiles has been continually rising. Car-mounted semiconductors come in a variety of varieties. More semiconductors are anticipated to be employed, particularly for ADAS (Advanced Driver Assistance Systems) in the future. Semiconductors enable us to live pleasant lives in this way.

## REFERENCES AND SUGGESTED READINGS

1. David.J.Griffiths, *Introduction to Quantum Mechanics,* Pearson Prentice Hall, 2005.
2. Chenming Calvin Hu, *Modern Semiconductor Devices for Integrated Circuits,* Pearson Prentice Hall, 2009.
3. S.O.KASAP, *Principle of Electronic Materials & Devices,* MC Graw Hill Education, 2018.
4. S.Datta, *Quantum Transport:Atom to Transistor,* Cambridge University Press, 2005.
5. Robert F. Pierret, *Semiconductor Device Fundamentals,* ADDISON-WESLEY Publishing Company, 1996.
6. Donald A. Neamen, *Semiconductor Physics and Devices Basic Principle,* MC Graw Hill Companies, 2012.
7. Ben G, Streetman, Sanjay Kumar Banerjee, *Solid State Electronic Devices,* Pearson Education Limited, 2016.
8. *https://physicsinmyview.com/2020/09/facts-about-quantum-physics.html*
9. *https://onlinecourses.nptel.ac.in/noc21_ee80/course*
10. *https://onlinecourses.nptel.ac.in/noc21_ee59/course*
11. *https://onlinecourses.nptel.ac.in/noc21_ee19/course*

# 02 SEMICONDUCTOR DIODES

## UNIT SPECIFICS

*Through this unit, we have discussed the following aspects:*

- *Describe how excess carriers are produced and recombined in a semiconductor*
- *Apply the Poisson and continuity equations to an analysis of the characteristics of excess carriers*
- *Identify the band diagram of a PN junction in thermal equilibrium and talk about how the p and n areas are depleted*
- *Create a small-signal model of the PN junction to be utilized in relating the device's small-signal currents and voltages*
- *Analyse the voltage breakdown characteristics of the diode*
- *Discuss the basic principles of optical devices such as photodiodes and Solar cells*
- *Describe and analyze the Light Emitting Diode's fundamental mode of functioning (LED)*

*The practical applications of the topics are discussed to generate further curiosity, creativity and improve problem-solving capacity.*

*Besides giving a large number of multiple-choice questions as well as questions of short and long answer types marked in two categories following the lower and higher order of Bloom's taxonomy, assignments through several numerical problems, a list of references, and suggested readings are given in the unit so that one can go through them for practice. It is important to note that for more information on various topics of interest, some QR codes have been provided in different sections, which can be scanned for relevant supportive knowledge.*

*After the related practicals, based on the content, there is a "Know More" section. This section has been carefully designed so that the supplementary information provided in this part becomes beneficial for the users of the book. This section mainly highlights the initial activity, examples of some interesting facts, analogies, history of the development of the subject focusing on the salient observations and findings, timelines starting from the development of the concerned topics up to the recent time, applications of the subject matter for our day-to-day real life or/and industrial applications on a variety of aspects, and case study related to environmental, sustainability, social and ethical issues whichever applicable, and finally inquisitiveness and curiosity topics of the unit.*

## RATIONALE

*This chapter discusses how the nonequilibrium concentration electrons and holes behave about time and geographical coordinates. Many carriers control the electrical characteristics of semiconductor material;*

*their behaviour is examined to understand better how semiconductor devices work. A p and n semiconductor are in contact to create a PN junction. The situation is similar to the one described above. These topics are covered in a non-equilibrium case. The expansion of several related issues are made explicit to assist readers in having a clearer understanding of the issues underlying semiconductor diodes.*

*Since the properties and operation of these PN junctions are tightly connected, the initial focus is placed heavily on this essential element. The electrostatics of pn junctions with zero and reverse bias are covered in this chapter. Additionally, the convention of electrical signals into optical signals and optical energy into electrical energy using semiconductor materials. This chapter also covers the device's optoelectrical features.*

## PRE-REQUISITES

*Basics of Semiconductor Physics*

## UNIT OUTCOMES

*The list of outcomes of this unit is as follows:*

*U2-O1: Describe the principle of electron-hole pair generation and recombination*

*U2-O2: Describe the I-V characteristics of the PN junction diode*

*U2-O3: Explain the breakdown mechanism of semiconductor diode and load line analysis*

*U2-O4: Realize the nonequilibrium behaviour of the Zener diode*

*U2-O5: Develop the optoelectronics devices such as LED, Photodiode, and Solar cells.*

| Unit-2 Outcomes | EXPECTED MAPPING WITH COURSE OUTCOMES (1- Weak Correlation; 2- Medium Correlation; 3- Strong Correlation) | | | | | |
|---|---|---|---|---|---|---|
| | CO-1 | CO-2 | CO-3 | CO-4 | CO-5 | CO-6 |
| U2-01 | 3 | 3 | 3 | - | 3 | 1 |
| U2-02 | 1 | 1 | 2 | 2 | 1 | - |
| U2-03 | 2 | 1 | 3 | 1 | 2 | 1 |
| U2-04 | - | - | 3 | 1 | 2 | 2 |
| U2-05 | 3 | 3 | 3 | - | 3 | 1 |

## 2.1 GENERATION AND RECOMBINATION OF CARRIERS

This section is about the mechanism of carrier generation and recombination. So, firstly, what do we mean by generation and recombination? By definition, generation implies that the electron and hole pairs are created, and recombination is the process by which the electron and holes are eliminated. The electron and hole concentration in semiconductors tend to fluctuate if a thermal equilibrium is disturbed. For instance, a fast temperature rise will speed up the thermal generation of electrons and holes, causing their concentrations to fluctuate over time until a new equilibrium value is established.

A non-equilibrium situation is caused by stimulation. For example, a photon flux may also generate electrons and holes. We shall begin by direct band-to-band creation and recombination to comprehend the processes. Later, the impact of permitted electron energy levels inside the band gap is referred to as recombination hubs or traps.

## 2.1.1 Semiconductor in Equilibrium

At thermal equilibrium, concentration of electrons in the conduction band and holes in the valence band have been determined. Such concentrations are not time-dependent in thermal equilibrium. The random nature of the thermal process causes electrons to be thermally stimulated continuously from the valence band into the conduction band. At the same time, the conduction band electrons randomly travelling across the crystal may hit holes and 'jump' into the valence band's empty states. The electron and hole are both annihilated during this recombination process.

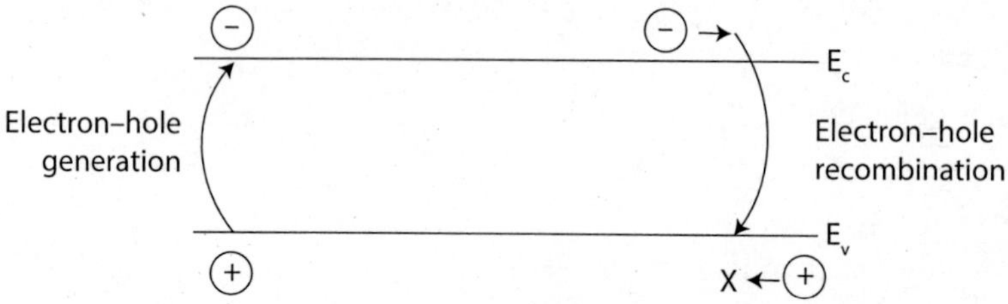


**Fig. 2.1:** Representation of Generation and Recombination

The creation and recombination of electrons and holes must occur at equal rates because the net carrier concentrations at thermal equilibrium are independent of time. In Fig. 2.1, the generation and recombination processes are depicted schematically.

Consider the thermal-generation rates of electrons and holes are $G_{no}$ and $G_{po}$. The holes and electrons are produced in pairs for direct band-to-band creation. Therefore, the generation rate must be equal.

$$G_{no} = G_{po} \tag{2.1}$$

Then, the recombination of electrons and holes is considered as $R_{no}$ and $R_{po}$ for the semiconductor equilibrium since electrons and holes recombine in pairs during direct band-to-band recombination.

The recombination rate for electron and hole is expressed as,

$$R_{no} = R_{po} \tag{2.2}$$

The generation and recombination rates are similar because the electron and hole concentration in thermal equilibrium are time-independent parameters. So, we have

$$G_{no} = G_{po} = R_{po} = R_{po} \tag{2.3}$$

## 2.1.2 Excess Carrier Generation and Recombination

A list of the most important symbols that are used in this chapter is shown in Table 2.1.

**Table 2.1:** Relevant notations used in this chapter

| Symbols | Definition |
|---|---|
| $n_o$, $p_o$ | Electrons/holes concentrations at thermal equilibrium |
| n, p | Concentrations of all electrons/holes (which may depend on place or time) |
| $\delta_n = n - n_o$ | An excess concentration of electrons |

| $\delta_p = p - n_p$ | Excess hole concentrations |
|---|---|
| $g_n', g_p'$ | Excess rates of electron/hole generation |
| $R_n', R_p'$ | Excess rates of electron/hole recombination |
| $\tau_{no}, \tau_{po}$ | An excess lifetime of the electron/hole in minority carriers |

When high-energy photons are shined on a semiconductor, valence band electrons could be stimulated to the conduction band. This causes the formation of an electron-hole pair. The additional electrons and holes are known as excess electrons and excess holes. Excess carriers may be produced in a semiconductor when it is out of equilibrium. If the concentration of carriers is in thermal equilibrium and in non-equilibrium, then . There are essentially three different generations, and they are as follows:

i. **Photon Generation**

When the light strikes a semiconductor during photon generation, if the energy of the light () is greater than the bandgap of the semiconductor, then an electron-hole pair will be formed and electron moves towards the conduction band.

Fig. 2.2 demonstrates how electrons traverse the conduction band after being struck by light with energy (hv > $E_g$), leaving a hole. It can accept an electron in higher conduction band states for various light wavelengths with various energies ($hv_2$, $hv_3$). Assume that a low-doping thin semiconductor plate is coupled to a voltage source and light is directed towards the plate. We obtain varied currents in the external circuit for various wavelengths, and this graph will resemble that in Fig. 2.3 (b).

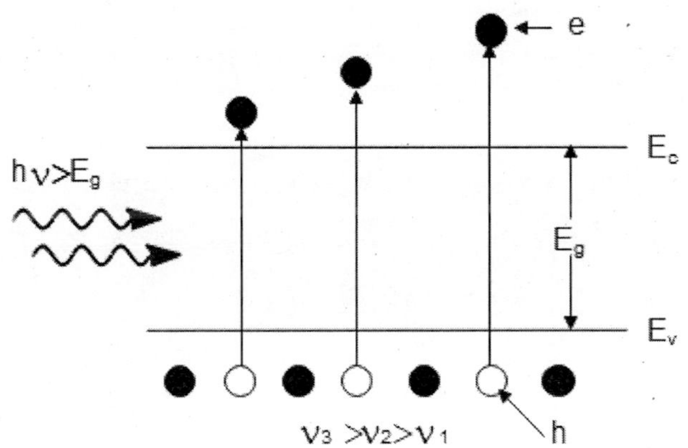


**Fig. 2.2:** Photon Generation

When the light frequency exceeds the critical frequency (v), a minimal amount of current, known as dark current, flows through the semiconductor. Then, as a result of the applied field, electron-hole pairs are created and begin to conduct. Therefore, the current grows quickly. Higher frequencies will result in constant current since the generation rate, which depends on the number of photons in light, will also stay constant. The generation rate will rise as light intensity increases (the number of photons increases). Current increases as a result. The relationship between current and light frequency ($I_1$>$I_2$>$I_3$) is seen in Fig. 2.3 (b).

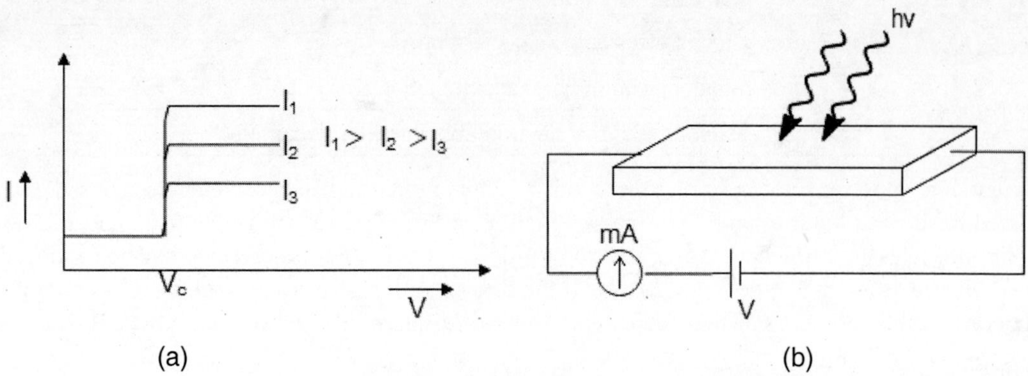


**Fig. 2.3:** Photon Generation: (a) Experimental setup,
(b) Current Vs frequency plot for different intensities of light

ii. **Phonon Generation**

When a semiconductor is overheated, phonon production takes place. More phonons are produced when the semiconductor's temperature rises due to increased lattice vibrations. Covalent bonds in the semiconductor dissolve due to increased lattice vibrations, creating electron-hole pairs.

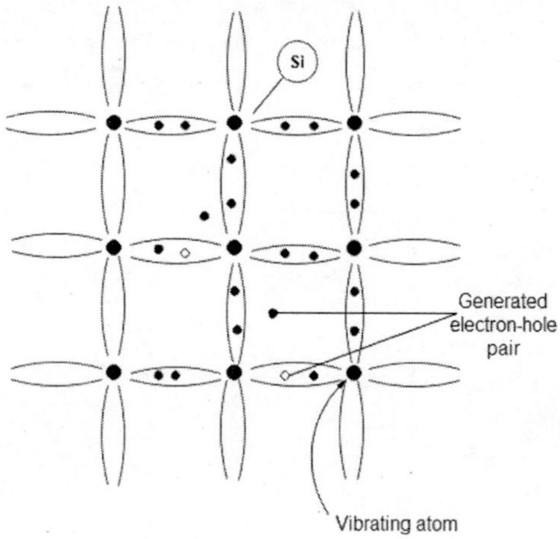


**Fig. 2.4:** Phonon Generation

iii. **Impact Ionization**

One energetic charge carrier will generate another charge carrier throughout this procedure. Semiconductors under an induced field cause electrons to gain high energy and strike additional Si-atoms between two subsequent collisions. A bond breaks out during this process, creating new carriers. Avalanche breakdown happens for very high electric fields.

A pair of electrons and holes get annihilated during recombination. The fact that, a conduction band free electron moves towards the valence band and unites with a hole to form a bound electron in the valence band does not indicate that they are destroyed.

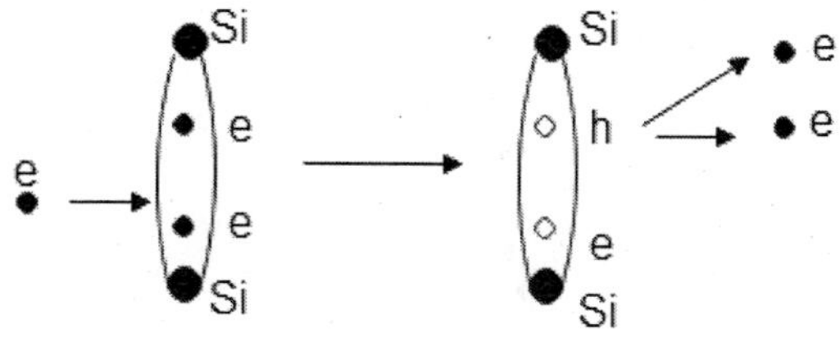


**Fig. 2.5:** Impact Ionization

### 2.1.3 Radiative Recombination

This process results in one energetic photon ($E_g$) emission, when the carrier from the conduction band minimum drops to the valence band maximum without change in momentum.

When stimulated to higher energy levels, electrons enter the conduction band and release their energy as heat. It is sometimes referred to as direct recombination. This process usually occurs in direct band gap semiconductors like GaAs.

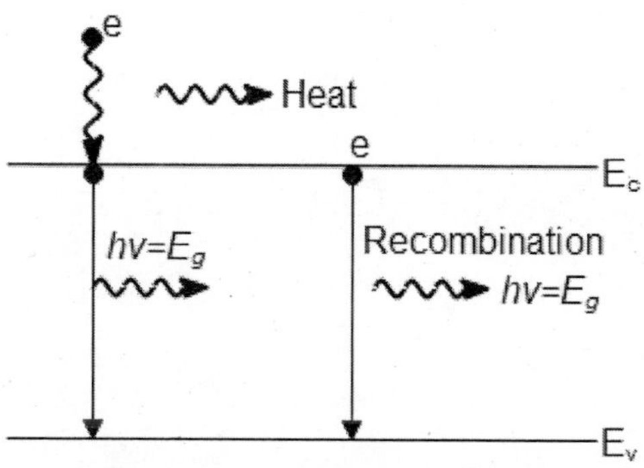


**Fig. 2.6:** Radiative Recombination

### Shockley-Read-Hall Recombination

In this recombination, electrons from the conduction band minimum travel to a defect level intermediate between Ec and Ev by emitting energy as photons or phonons, and then travel to the valance band from that intermediate level. This form of recombination is observed in impure semiconductors and semiconductors containing defects. Commonly, the level of faults is in the centre of the forbidden gap.

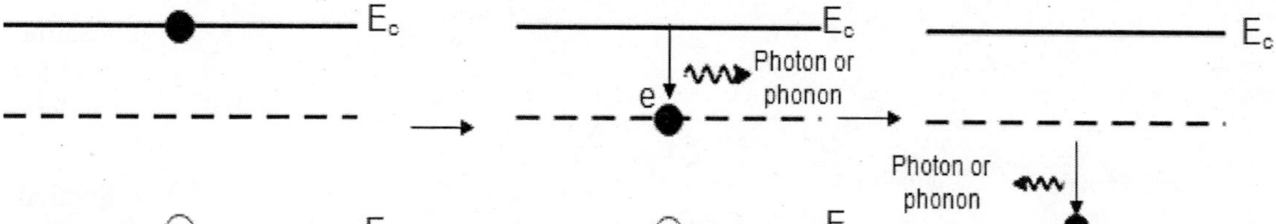


**Fig. 2.7:** S-R-H Recombination Process

## AUGER RECOMBINATION

There are three carriers in Auger recombination. In this process, an electron-hole recombines, but the energy is transferred to the third free electron in the conduction band rather than discharging it as a photon or phonon. The third exciting electron then returns to the conduction band edge by producing heat as it does so. For strongly doped material, an Auger recombination often takes place.

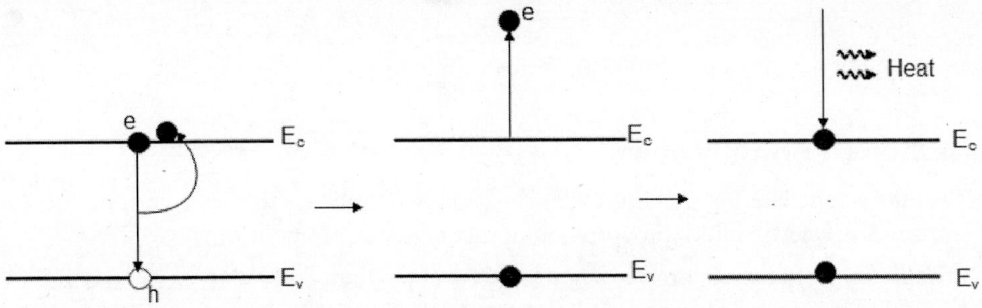


**Fig. 2.8:** Auger Recombination Process

According to the semiconductor, the recombination process may also be divided into two categories: direct and indirect recombination.

### Direct Recombination

Direct recombination occurs when the valence band's maximum and conduction band's minimum energy levels occur for a similar momentum. When electrons move from the conduction band edge directly hitting the valence band, they generate photons with an energy equivalent to the semiconductor's bandgap.

### Indirect Recombination

However, in indirect semiconductor recombination, electrons from the conduction band must shift momentum to approach the valence band top, at which point recombination occurs. Indirect recombination often involves heat being released and absorbed by the semiconductor.

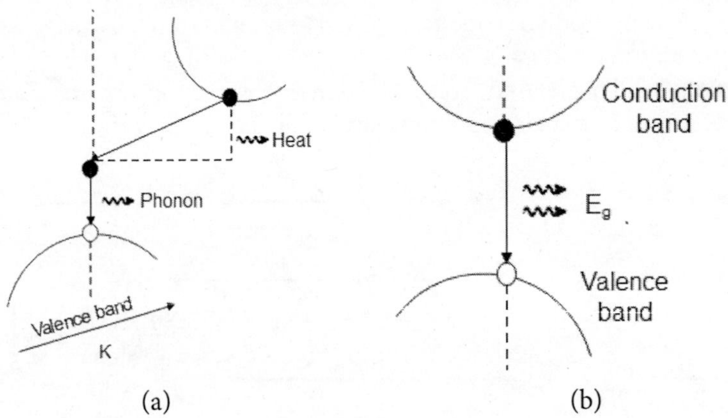


**Fig. 2.9:** (a) Direct Recombination, (b) Indirect Recombination

External force produces the extra electrons and holes at a certain rate. Let be the rate of extra electrons produced, and $g'_p$ be the rate of extra holes produced. Additionally, these generation rates have units of $cm^{-3}$ $s^{-1}$. The excess electrons and holes are produced in pairs for direct band-to-band creation, therefore we need to have,

$$gn' = gp' \tag{2.4}$$

More electrons and holes are produced above thermal equilibrium. We may write the excess carrier generation as,

$$n = n_o + \delta_n \tag{2.5 (a)}$$

$$p = p_o + \delta_p \tag{2.5 (b)}$$

where $\delta_n$ and $\delta_p$ are excess electron and hole concentrations, respectively, and $n_o$ and $n_p$ are the thermal equilibrium concentrations.

The method of producing excess electron holes is depicted in Fig. **2.9**. The semiconductor is not in thermal balance as a result of the external force's perturbation of the equilibrium condition. From equation 2.5 (a) and 2.5 **b** we may deduce that, under non-equilibrium conditions,

$$np \neq n_o p_o = n_i^2 \tag{2.6}$$

The carrier concentrations steadily increase due to a steady-state emission of more electrons and holes. Conduction band electrons have the potential to "slip down" into the valence band, initiating the process of excess electron-hole recombination. and are recombined electron and hole respectively. Recombination must be equal since extra holes and electrons combine in pairs. So, we can express the recombination rate as,

$$R'_n = R'_p \tag{2.7}$$

In direct band to band recombination, the recombination occurs spontaneously, thus, the probability of an electron and hole recombining is constant at time. Both the electron and hole concentrations should be proportional to the rate of electron recombination. Recombination is impossible if there are no electrons or holes. The net rate of increase in the carrier density can be expressed as,

$$\frac{dn(t)}{dt} = \alpha_r[n_i^2 - n(t)p(t)] \tag{2.8}$$

$$n(t) = n_o + \delta_n (t)$$

$$p(t) = p_o + \delta_p (t)$$

The first term in the equation 2.8 $\alpha_r$ $n_i^2$ represents the thermal equilibrium generation rate. We know that $\delta_n (t) = \delta_p (t)$ because extra electrons and holes are generated and pairwise recombine. If $n_o$, $p_o$ are time-independent thermal equilibrium parameters, then the equation becomes,

$$\frac{d(\partial_n(t))}{dt} = \alpha_r[n_i^2 - (n_o + \partial_n(t))(p_o + \partial_p(t))]$$

$$= -\alpha_r \delta_n (t)[(n_o + p_o) + \delta_n (t)] \tag{2.9}$$

If we enforce the requirement of low-level injection, equation 2.9 can be solved with ease. The amount of the concentration of additional carriers relative to the steady state carrier concentrations at room temperature is constrained by low-level injection. We often see $n_o \gg p_o$ in extrinsic n-type elements and $p_o \gg n_o$ in extrinsic p-type elements. When there is low-level injection, the majority carrier concentration at thermal equilibrium is substantially lower than the excess carrier concentration. On the other hand, high level injection happens when the additional carrier density equals or exceeds the majority carrier's concentrations at thermal equilibrium.

When a p-type material is taken into account under low-level injection ($\delta_n$ (t) ≪ $p_o$), equation 2.9 changes,

$$\frac{d(\partial_n(t))}{dt} = -\alpha_r p_o \partial_n(t) \tag{2.10}$$

The above equation can be written as an exponential decay function from initial excess concentration.

$$\partial_n(t) = \partial_n(0)e^{-\alpha_r p_o t} = \partial_n(0)e^{-\frac{t}{\tau no}} \tag{2.11}$$

where the constant for low-level injection is $\tau_{no}$ = $(\alpha_r p_o)^{-1}$. $\tau_{no}$ is commonly known as the extra minority carrier lifespan since equation (2.11) explains how additional minority carrier electrons decay.

Equation 2.11 may be used to express the recombination rate of additional minority carrier electrons, which is expressed as a positive value as,

$$R'_n = -\frac{d(\delta_n(t))}{dt} = +\alpha_r p_o \delta_n(t) = \partial_n(t)/\tau_n o \tag{2.12}$$

The additional majority carrier holes merge at the same rate for direct band-to-band recombination, so for the p-type material,

$$R'_n = R'_p = \delta_n(t)/\tau_{no} \tag{2.13}$$

The decay of minority carrier holes happens in an n type material ($n_o$ ≫ $p_o$) under the circumstance of low-level injection ($\delta_n$ (t) ≪ $n_o$) having a time constant of, $\tau_{po}$ = $(\alpha_r\ n_o)^{-1}$ we have,

$$R'_n = R'_p = \delta_n(t)/\tau_{no} \tag{2.14}$$

## 2.2 POISSON AND CONTINUITY EQUATION

The Poisson's equation and the continuity equation are playing an important role in macroscopic semiconductor device simulation. One of the fundamental equations in electrostatics is Poisson's equation, which is derived from Maxwell's equation and the material relation ∇.D= ρ. The electric displacement field is denoted by D = ϵE, the electric field by, the charge density by ρ, and the permittivity tensor by $\hat{\epsilon}$. Poisson's equation is obtained by combining the electrostatic potential ψ with E = -∇ψ as,

$$\nabla.(\hat{\epsilon}\ \nabla\ \psi) = -\rho \tag{2.15}$$

For high frequencies, when wavelength is much less than the device dimension, the quasi-stationary approximation used to verify Poisson's equation is still applicable. In this derivation, the permittivity tensor is also considered time invariant. The scalar value can also be used to approximate permittivity in isotropic materials like silicon. Furthermore, within a material section, permittivity is frequently assumed to be constant. The charge density in semiconductors is generally divided into fixed charges, such as ionized acceptors and donors , and free charges, such as electrons and holes . Poisson's equation can be written as if acceptors, donors, electrons, and holes are included.

$$\nabla^2\psi = \frac{q}{\in}(n - p + N_A^- - N_D^+) \tag{2.16}$$

In ideal oxides, the right side of equation 2.16 becomes 0, and Poisson's equation changes to the Laplace equation. However, it is necessary to consider the presence of oxide charges $N_{ox}$, which produces in,

$$\nabla^2\psi = \frac{q}{\in} N_{ox} \qquad (2.17)$$

The interface trap concentration must also be considered at interfaces and surfaces. Maxwell's equations can also be used to derive the continuity equation, which gives,

$$\nabla.J + \frac{\partial p}{\partial t} = 0 \qquad (2.18)$$

For the contribution of electrons and holes, the current density is divided into and . The elementary charge is represented by , and the concentrations of electron and hole are expressed as and respectively. Two different continuity equations can be formed by introducing the recombination rate as,

$$\nabla.J_n - q\,\frac{\partial p}{\partial t} = +\,qR(\text{For electrons}) \qquad (2.19)\ (a)$$

$$\nabla.J_n - q\,\frac{\partial p}{\partial t} = +\,qR(\text{For holes}) \qquad (2.19)\ (b)$$

The splitting of the equations into two, allows for independent carrier-type transport modelling. The rate denotes the rate of change, the value of *R* will be zero at thermal equilibrium, implying that generation and recombination are equal. The electron-hole pair generation and recombination are expressed using recombination models that are either physically or empirically based. The equations 2.15 and 2.16 can also be derived using the method of moments from Boltzmann's transport equation.

## 2.3 P-N JUNCTION CHARACTERISTICS

PN junction is actually a junction created between n and p type semiconductor as shown in Fig. 2.10. Each region has both types of charge carriers. A metallurgical junction refers to the boundary between two sections at x=0.

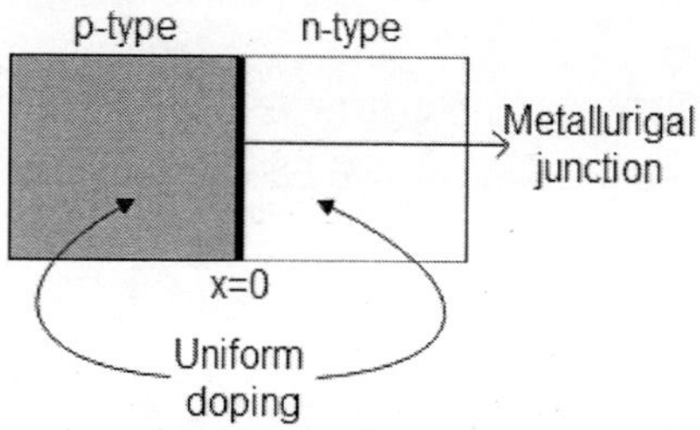


**Fig. 2.10:** PN Junction

### 2.3.1 Theory of P-N Junction

For the sake of simplicity, let's take a closer look at a step junction with uniform doping concentrations throughout the entire region and a sudden change in doping at the junction is shown in Fig. 2.11. At thermal equilibrium or when no external bias is applied, the number of holes and electrons are different on each side of a junction. So, the holes from the p-side can diffuse to the n-side, and electrons from the n-side can diffuse to the p-side. They cause a flow of diffusion current through the junction. Also, an immobile, ionised donor is left on the n-side. As the process proceeds, a layer of positive charge forms on the junction's n-side. An ionised acceptor is left behind when a hole moves from the p-side to the n-side.

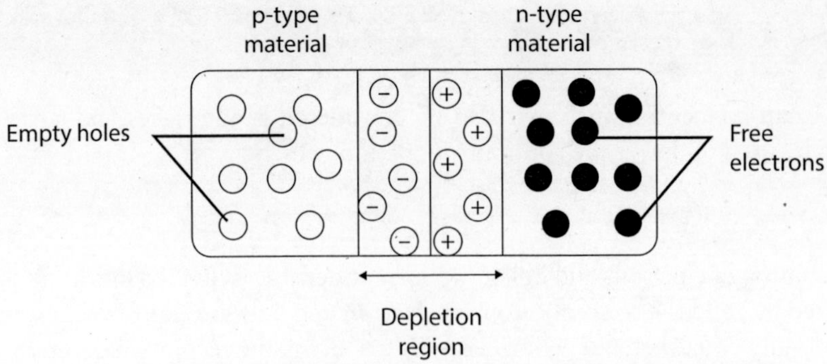


**Fig. 2.11:** Representation of Depletion Region

The junction near the p-side develops a layer of negative charges as a consequence. The area around the junction which is depleted of the free charge carriers is known as depletion region. We can classify the PN junction into four categories based on their doping concentration.

1. Abrupt junction
2. Linearly graded junction
3. $P^+$ N junction
4. PIN junction

**Abrupt Junction** - When the electron concentration is higher than the hole concentration or the concentration of hole is higher than the electron, there is a sudden change in the doping profile. These sudden changes appear as the step type changes at the junction. This kind of doping profile is called as step graded PN junction or an abrupt p-n junction. Example: Varactor diode.

**Linearly Graded Junction** - In the entire region of a p-n junction, the doping concentration shifts linearly from n-type to p-type and vice versa. This kind of PN junction is known as a linearly graded junction.

**P+N junction** - This kind of junction is doped heavily at the p-type region compared to the n-type.

**PIN junction** - The intrinsic region is sandwiched between the P and N regions.

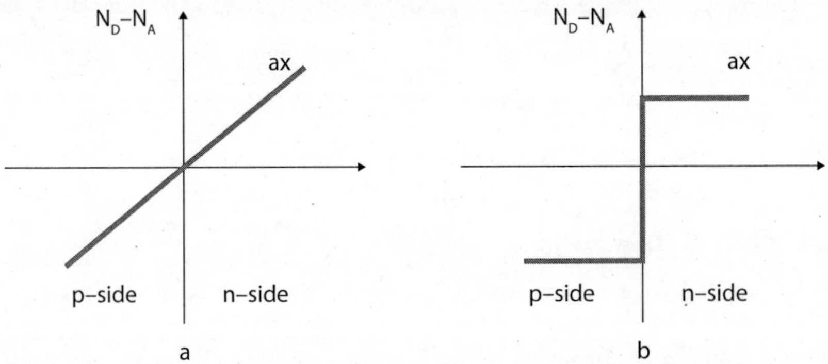


**Fig. 2.12:** a) Abrupt Junction, b) Linearly graded Junction

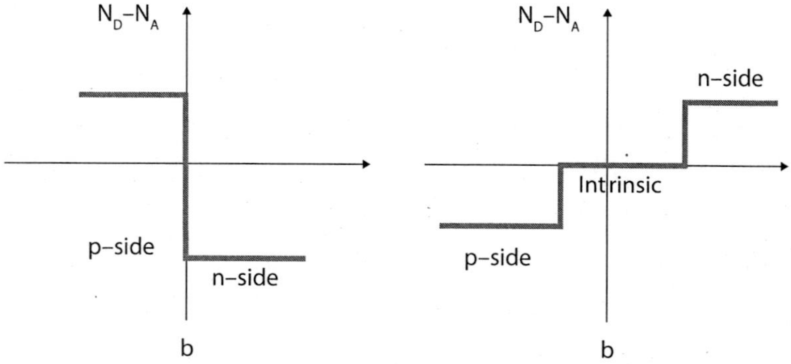


**Fig. 2.13:** a) PIN Junction, b) P+N Junction

### 2.3.2 Energy Band Diagram of P-N Junction

The fermi level is located just above the valence band for p-type materials and just below the conduction band for n-type materials. Diffusion begins when the p-n junction forms. The concentration of charge carriers changes due to the movement of charge carriers from one semiconductor material to another semiconductor material until the fermi level of the two materials aligns itself. This is comparable to adjusting the water levels in two linked tanks when they are at different levels. Until the fermi levels on the two sides line up, charge flows from p to n and n to p side. Fig. 2.14 illustrates the energy band structure and charge transfer.

In p-region, the fermi level ($E_F$) is near the edge of the valence band ($E_V$) and in n region, the fermi level is near the edge of conduction band ($E_C$). In p-type materials, the edge of the conduction band is higher than in n-type materials due to the transfer of charges. P-type materials have a greater valence band edge than n-type materials. As a result, the intrinsic levels of the fermi level on both side shifts.

To achieve a comparable fermi level for the p-n junction, the n and p side fermi levels are adjusted. The contact difference in potential across the junction is caused by the total energy shift $E_0$, which is made up of the components $E_1$ and $E_2$. This is simply the barrier potential, junction potential, or contact potential provided by,

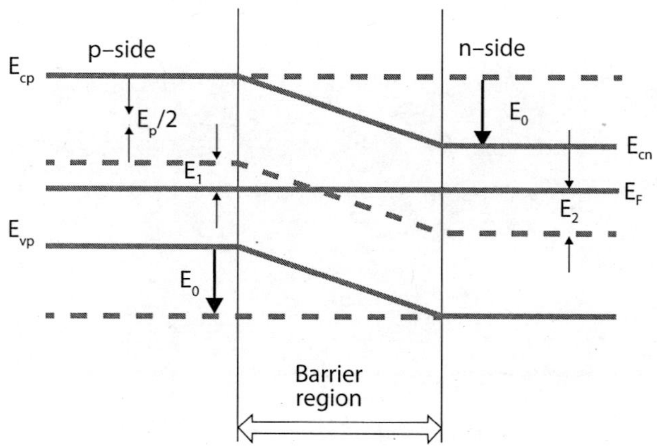


**Fig. 2.14:** Band Diagram of p-n Junction

$V^{bi} = KT \ln (N_A \; N_D/n_i^2$ (2.20)

where, $N_A$ = Concentration of acceptor impurity

$N_D$ = Concentration of donor impurity

$n_i$ = Intrinsic concentration

k = Boltzmann's constant = $8.61\times10^{-5}$ $e^{V/K}$

T = Temperature (Room temperature = 300°C)

### 2.3.3 Biasing of P-N Junction

A P-N junction is biased by connecting an external voltage source across its ends. Depending upon the way of connection, there can be different scenarios as follows,

**Zero Bias Configuration:** The P-N junction is not being charged externally.

**Forward Bias Configuration:** Positive and negative terminals are connected to the p-type and n-type, respectively.

**Reverse Bias Configuration:** The p-type is connected to the -ve terminal, therefore the n-type is linked to the +ve terminal.

**Forward Bias Configuration**

In forward bias, a positive potential difference is applied to PN junction by connecting a p-type semiconductor to positive terminal of the battery and a n-type semiconductor to negative terminal of the battery.

The negative electrode of the battery drives the free-electrons against the potential barrier from the n to the p-region, when the supplied voltage exceeds the barrier potential. In a similar manner, the positive voltage pushes the holes towards n-region. Since the holes are repelled by the positive potential and try to cross over the barrier. The applied electric field at the P-N junction is pointed in the reverse direction as the induced electric field. The combined electric field has a lesser magnitude than the induced electric field when the two electric fields are added up.

As a result of this, the depletion area becomes less resistive and thinner. When there is a lot of applied voltage, the resistance of the depletion area is very low. At 0.6 V in silicon, the depletion region's resistance totally vanishes, allowing current to pass through it unhindered.

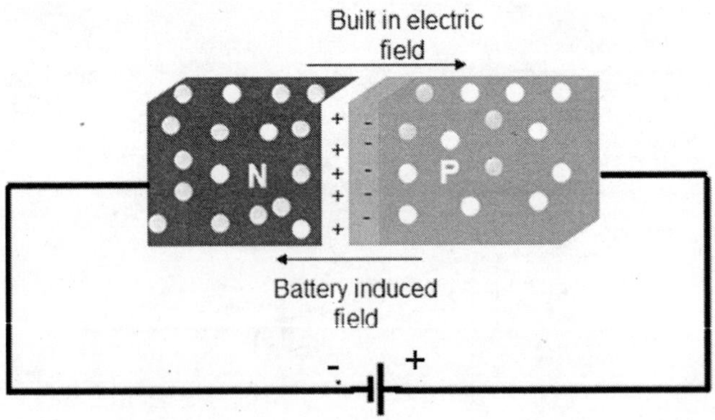


**Fig. 2.15:** Forward Biased p-n Junction

### Reverse Bias Configuration

In this configuration, the negative electrode pulls the holes in the p-side away from the junction. The positive electrode draws the unbound n-region electrons away from the junction. No charge carrier could pass the intersection. The depletion zone expands as the electron and hole travel away from the junction. The production of more positive ions raises the positive charge in the n-region, while the production of more -veions raise the negative charge in the p-region. This is due to the barrier potential being aided by the supplied voltage. In this case, both the built-in and applied electric fields are pointing in the same direction. A thicker, more resistive depletion area results from the combination of the two electric fields because the resultant electric field is directed in the same direction as the induced electric field. The depletion area thickens and becomes more resistive as the applied voltage rises.

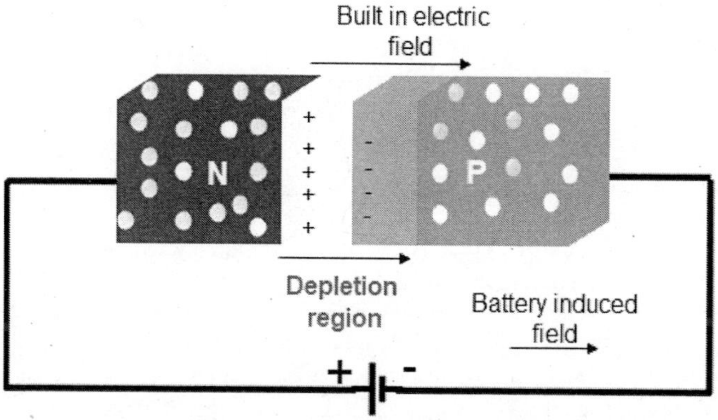


**Fig. 2.16:** Reverse Biased p-n Junction

## 2.3.4 I-V Characteristics of P-N Junction

The P-N junction diode is in *zero-bias* condition, when no external voltage is applied between its terminals. In this condition, it blocks current conduction due to the barrier potential at the junction.

When the diode is biased forward, the current progressively increases, and when the voltage is applied over the barrier potential, a non-linear curve is produced. The diode continues to function correctly once it has passed the potential barrier. As the external voltage rises, the curve rises quickly.

When the p-type material is connected to the negative terminal of the external battery while the n-type material is connected to the positive terminal then the P-N junction diode is in a *negative bias* condition. The potential barrier rises as a result of this. Minority carriers are present at the connection at first, causing reverse saturation current to flow.

The energy of minority carriers grows as negative bias applied across the P-N junction increases and beyond a certain potential this provides sufficient energy to break free electrons from valence band of other atoms. This results in positive feedback which increases current substantially and make the diode useless. The results are shown in Fig. 2.17.

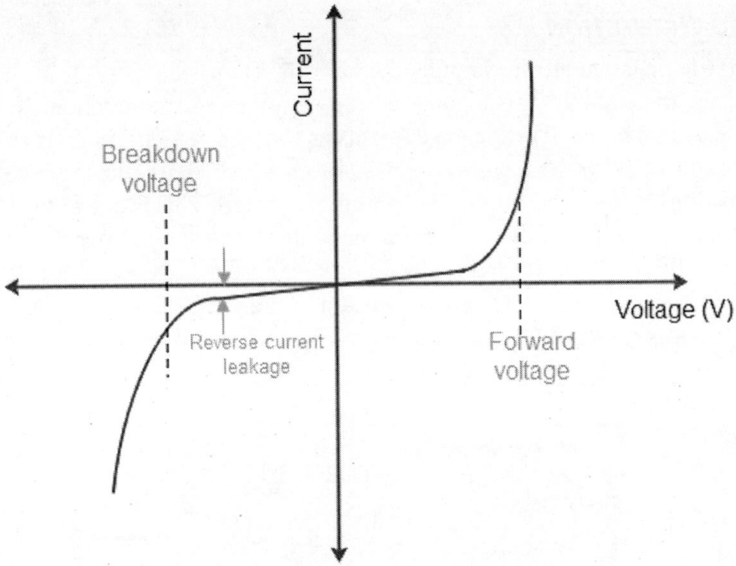


**Fig. 2.17:** I-V Characteristics

**Example 2.1:** Determine the built in potential of the pn junction having doping concentration of $N_a = N_d = 10^{10}$ cm$^{-3}$, ($n_i = 1.5 \times 10^{10}$ cm$^{-3}$). Assume room temperature.

**Given:** Doping and acceptor concentration of $N_a = N_d = 10_{10}$ cm$^{-3}$

**Solution:**

Form the equation 2.20, we can calculate the built-in potential as,

$$V_{bi} = KT \ In\left(NA\frac{ND}{n_i^2}\right) = 8.61 \times 10^{-5} \times 300 \ In\left(NA\frac{ND}{n_i^2}\right)$$

$$V_{bi} = 0.0258 \ In\left(10^{10} \cdot \frac{10^{10}}{\left(1.5 \times 10^{10}\right)^2}\right)$$

$$V_{bi} = 0.021 \text{ V}$$

## 2.4 LOAD LINE ANALYSIS OF P-N JUNCTION

The load line analysis of the diode circuit illustrates the limits placed on other circuit components and aids in understanding the nonlinear characteristics of the circuits. Therefore, by utilising the load line, we can understand how the diode is constrained by other components of the circuit.

Let us take a P-N junction diode and connect it as illustrated in Fig. 2.18. The current flow into the circuit is represented by $I_D$. We can quickly analyse the circuit by using Kirchhoff's voltage law.

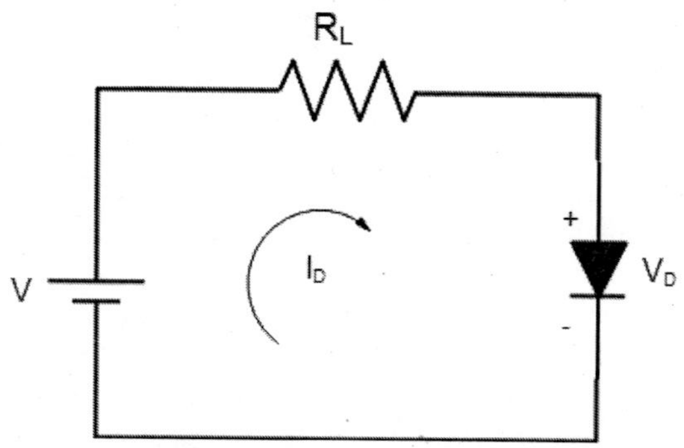


**Fig. 2.18:** Circuit Setup

$$V = I_D R_L + V_D \tag{2.21}$$

The equation 2.22 denotes the value of the current when the voltage across the diode is zero.

$$V = I_D R_L + 0$$

$$I_D = \frac{V}{R_L} \tag{2.22}$$

In the same way, is calculated by making as zero.

$$V = 0 + V_D$$

$$V_D = V \tag{2.23}$$

The two points (V/RL,0) & (0, V) from the equation is connected to form a straight line, which is referred to as the diode's "*load line*".

The intersection of diode characteristics and load line has a point known as the "Q point", sometimes known as an operating point or a quiescent point. Since the point is equal to the operational voltage (VDQ) and current (IDQ), we obtain

$$Q_{point} = (I_{DQ}, V_{DQ}) \tag{2.24}$$

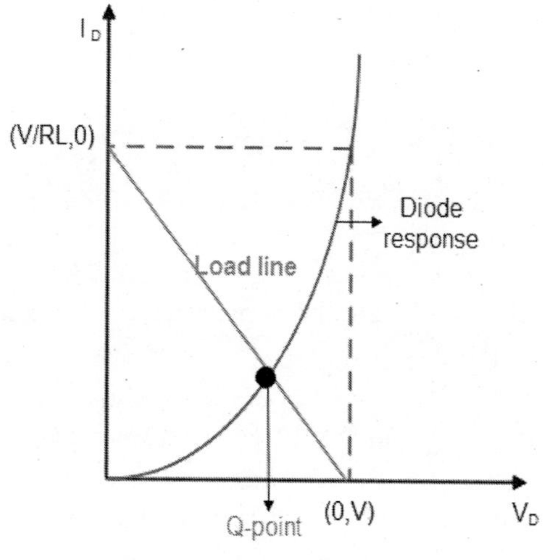


**Fig. 2.19:** Diode Load Line

Finally, the intersection of the load-line and diode characteristics is the *operational point*. The slope of the load line is found by dividing the equation 2.21 by $R_L$,

$$\frac{V}{R_L} I_D + \frac{V_D}{R_L} \text{ or } I_D = \frac{V_D}{R_L} + \frac{V}{R_L}$$

The above equation is same like a straight-line equation, the slope is negative,

$$-\frac{1}{R_L} \tag{2.25}$$

The Q-point is very sensitive to the parameter $R_L$. The Q-point will change if one change the value of $R_L$, the slope will also change.

**Example 2.2:** Find the Q-point for the circuit shown in Fig. 2.18 having $R_L = 10\Omega$ and applied voltage V=5V.

**Given:** Load resistance of $R_L = 10\Omega$ and V=5V.

**Solution:**

The Q-points are (IDQ, 0), (VDQ,0).

$$I_{DQ} = \frac{V}{R_L} = \frac{5}{10} = 0.5A$$

$$V_{DQ} = V = 5V$$

So, the points are (0.5A,0) and (5V,0) based on the points load line is drawn.

## 2.5 SMALL SIGNAL MODEL

Small signal model is a linear approximation of nonlinear device which is valid only for small signals.

### Steps

1. Determine the DC operating points ($V_D$ & $I_D$) of diode.
2. A diode is subjected to DC voltage ($V_D$), which is modelled by a battery, and time varying signal $V_d$ (t) is overlaid on DC voltage $V_D$.
3. When there is no signal present, the diode voltage is equal to $V_D$ and the diode will conduct a DC current $I_D$ is,

   $$I_D = I_s e^{\frac{V_D}{nV_T}}$$

   where $I_S$ (saturation current) and n are diode parameters.
4. n is the fitting parameter that lies within the range of 1 and 2 depending on diode's material and structure.
   - For n=1, diode made using standard IC fabrication process when operated under normal condition.
   - For n=2, diode available as discrete 2-terminal components.

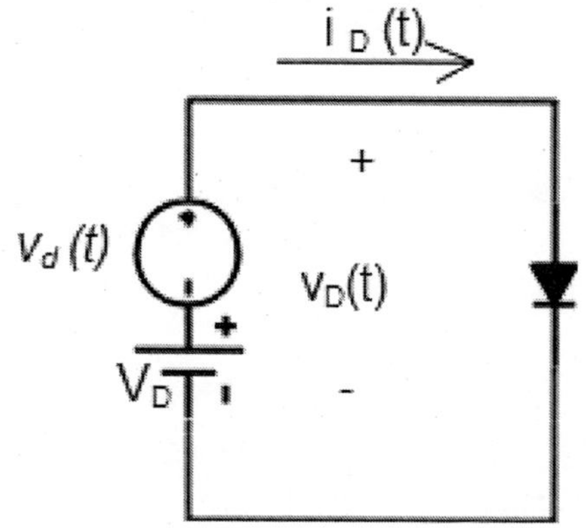


**Fig. 2.20:** Circuit for Small Signal Model

5. When $V_d$ (t) is applied, total instantaneous diode voltage, $V_D (t) = V_D + V_d (t)$ (2.26)

6. Total instantaneous diode current, $I_D = I_s e^{\frac{V_D}{nV_T}}$ (2.27)

7. Put equation 2.26 in 2.27, $i_D(t) = I_S e^{\frac{(V_D + v_d(t))}{nV_T}} = I_S e^{\frac{V_D}{nV_T}} e^{\frac{V_d(t)}{nV_T}}$

$$i_D(t) = I_D e^{\frac{V_d(t)}{nV_T}}$$

$$\text{where } I_D = I_S e^{\frac{V_D}{nV_T}}$$

8. If amplitude of signal $V_d$ (t) is kept sufficiently small, $\frac{Vd(t)}{nV_T} \ll 1$

9. So, expand exponential in series and truncate the series, small signal approximation of $i_D$ (t) reduced to,

$$i_D(t) = I_D\left(1 + \frac{v_d(t)}{nV_T}\right)$$

10. The above equation is suitable for signals which are having smaller amplitudes of 10mV for n = 2 and 5mV for n = 1, $V_T$ = 25mV[3].

$$i_D(t) = I_D + I_D\left(\frac{Vd(t)}{nV_T}\right)$$

$$i_d = I_D\left(\frac{Vd(t)}{nV_T}\right) \quad (2.28)$$

11. From the equation 2.28, it's clear that signal current and signal voltages both are in direct proportion.

12. Diode small signal conductance is the quantity that relates signal current $i_d$ to signal voltage $v_d$ and has dimensions of conductance (mho).
13. We can write the small signal resistance as,

$$r_d \alpha \frac{1}{I_D}$$

$$r_d = \frac{1}{\frac{i_D}{V_D}}, \text{ when } I_D = i_D \tag{2.29}$$

14. Small signal values $V_D$ (t) and $i_d$ (t), both having a common diode small signal resistance $r_d$ will be placed over quantities $V_D$ and $I_D$ that define DC *bias point* or *quiescent point* of diode.

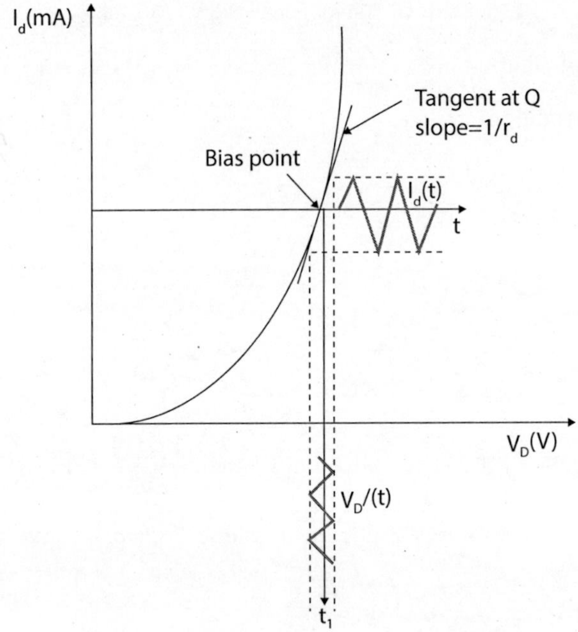


**Fig. 2.21:** Analysis of Small-Signal Diode Model

## 2.6 ZENER DIODE AND ITS CHARACTERISTICS

A heavily doped semiconductor device created to work in the reverse direction is a Zener Diode. It is usually referred to as a breakdown diode. A Zener diode's junction collapses and current flows in the opposite direction when the voltage between its terminals is reversed and the potential approaches the Zener Voltage (also known as the knee voltage). This effect is known as the *Zener Effect*. When current reaches the Zener voltage, diode allows the current to flow towards the cathode terminal, but also in the opposite direction.

### 2.6.1 Circuit Symbol of Zener Diode

The circuit symbol of a Zener diode is typically represented in Fig. 2.22. The symbol consists of a diagonal line that connects a horizontal line on one side (representing the anode) and a vertical line on the other side (representing the cathode).

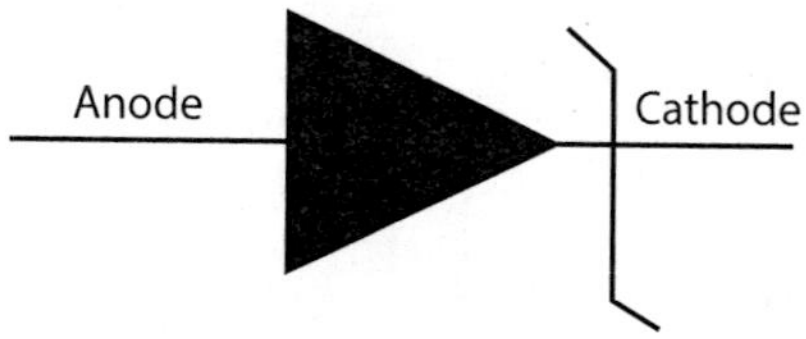


**Fig. 2.22:** Circuit Symbol of Zener Diode

### 2.6.2 Breakdown Mechanism of Zener Diode

When it is forward-biased, a Zener diode behaves just like a regular diode. However, when coupled in reverse biased mode, a little leakage current passes through the diode. Current begins to flow through the diode when the reverse voltage rises to the specified breakdown voltage ($V_z$). The series resistor determined the maximum current and the current remains constant throughout a broad range of applied voltage. We can classify the breakdown into two categories.

i. Avalanche Breakdown
ii. Zener Breakdown

#### *Avalanche Breakdown Mechanism*

The applied reverse potential is strong enough, avalanche breakdown happens in both Zener and regular diodes. The liberated electrons obtain enough energy and accelerate at high speeds when a higher reverse voltage is supplied to the PN junction. This speedy, free electron collides with other atoms resulting in the loss of additional electrons. The electric current in the diode increases quickly as a result of this, producing a significant number of free electrons. The conventional diode could be irreversibly damaged by this fast rise in electric current. A Zener diode can withstand the abrupt surge of current since it is made to function under avalanche breakdown.

#### *Zener Breakdown Mechanism*

The electric field in the depletion area becomes powerful and enough to remove electrons from the valence band as the applied reverse bias voltage approaches the Zener voltage. The parent atom is liberated by the valence electrons that have gained enough energy from the intense electric field of the depletion area. A slight change in voltage causes the electric current to grow quickly in the Zener breakdown zone.

### 2.6.3 Analysis of Zener Diode V-I Characteristics

The circuit setup for V-I characteristics of Zener diode is depicted in Fig. 2.23. The Zener diode's V-I characteristics is depicted in Fig. 2.24.

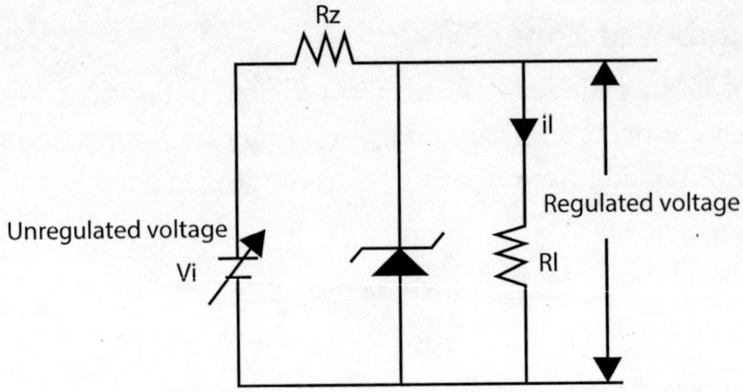


**Fig. 2.23:** Circuit Setup for V-I Characteristics of Zener Diode

A Zener diode's V-I properties can be classified into two categories as follows:

i) Analysis in forward mode

ii) Analysis in reverse mode

## Forward Mode of Zener Diode

The forward mechanism of a Zener diode is shown in the graph's top quadrant. We can infer from the graph that its forward properties are essentially the same as those of any other P-N junction diode.

## Reverse Mode of Zener Diode

A small reverse saturation current called $I_S$ flows across the diode when a reverse voltage is applied to a Zener diode. Minority carriers produced thermally are the cause of this current. At a given level of reverse voltage, the reverse current increases dramatically and abruptly as the reverse voltage rises. This reverse voltage, also known as Zener voltage, is designated by the symbol $V_Z$.

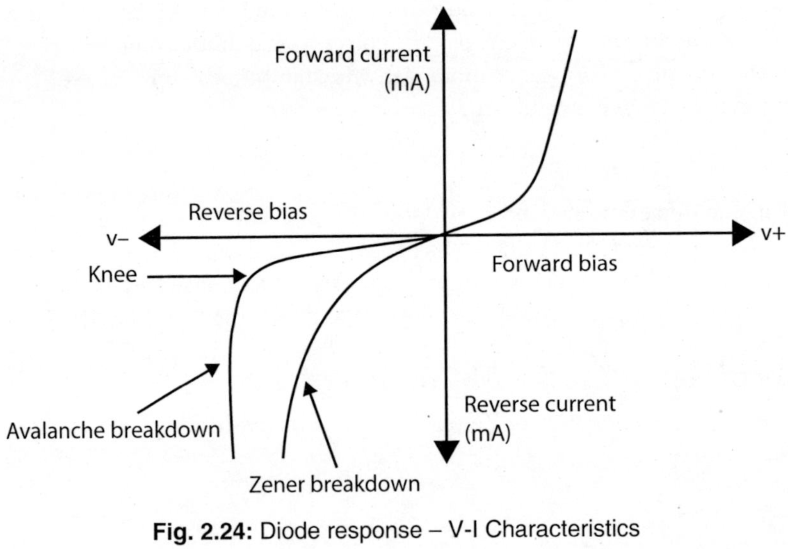


**Fig. 2.24:** Diode response – V-I Characteristics

### *Specifications for Zener Diode*

The following are some parameters for Zener diodes that are often used:

- **Zener Voltage/Breakdown Voltage** - The reverse breakdown voltage, also known as a Zener voltage, has a range of 2.4 V to 200 V, and it can occasionally reach 1 KV. The top voltage for a surface-mounted device is 47 V.
- **Current Iz (max)** – This is the highest current possible at the rated Zener voltage (Vz – 200 μA to 200 A).
- **Current Iz (min)** – This is the bare minimum current needed for a diode to break.
- The Zener diode's maximum permissible power dissipation is indicated by its **power rating**. It is determined by dividing the diode's voltage by the current passing through it.
- Diodes around 5V have the highest stability at high temperatures.
- **Voltage Tolerance:** Typically, it is 5%.
- **Zener Resistance (Rz)** - This refers to the Zener diode's resistance.

**DO YOU KNOW**

The device has the name of American physicist Clarence Zener, who initially identified the Zener effect in 1934 while conducting experiments on the breakdown of electrical insulator qualities that were mostly theoretical in nature. Later, Bell Labs used his work to create the Zener diode, an electronic device that mimics the phenomenon.

## 2.7 SCHOTTKY DIODE

The metal-semiconductor diode was one of the earliest devices used in the early 1900s. This type of diode, also known as a point-contact diode, by rubbing a metallic whisker across a semiconductor surface that was exposed. However, they were difficult to replicate and lacked mechanical reliability, pn junctions took their place in the 1950s. However, the production of repeatable and dependable contact, or Schottky barrier diodes, is now accomplished using semiconductor and vacuum technologies. We'll focus on n-type semiconductor diodes as they often have rectifying contacts constructed of this material.

The Schottky diode is in the form of metal-semiconductor junction diode. It is often referred to as the low voltage diode, Schottky barrier diode or hot-carrier diode The intersection of a semiconductor and a metal result in the formation of a Schottky diode. The Schottky diode features a low forward voltage drop and quick switching action. As is well known, a PN junction is created when p and n-type semiconductors are combined. In contrast, metals like platinum or aluminium are used in Schottky diodes in place of P type semiconductors.

### 2.7.1 Construction and Band Diagram

The schematic for a Schottky diode, which combines a metal with an n-type silicon semiconductor that has undergone light doping, is shown in Fig. 2.25 (a). This junction is known as a "metal-semiconductor junction".

This kind of metal compound and semiconductor material used in this metal-semiconductor junction's construction will have a significant impact on its width and, consequently, its electrical characteristics. However, electrons migrate from the n-type material to the metal electrode when the device is forward-biased, allowing current to flow. Therefore, the majority carriers drift produces the current that passes through the Schottky diode.

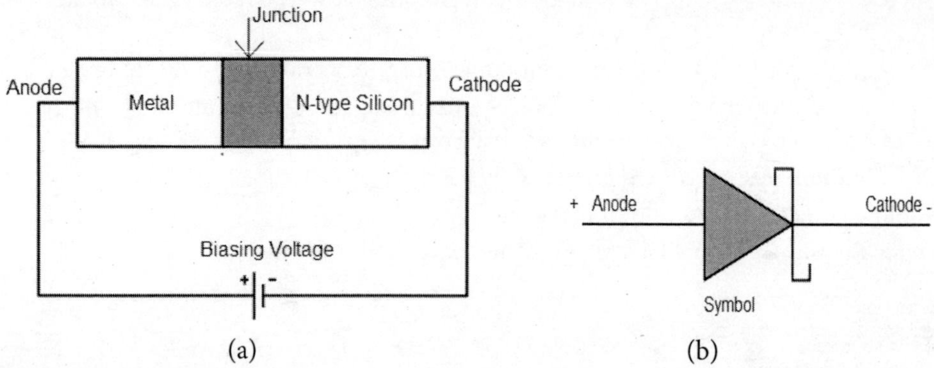


**Fig. 2.25:** Schottky Diode: a) Schematic of Schottky Diode, b) Circuit Symbol of Schottky Diode

When reverse biased, the diode's conduction abruptly ceases and transforms into blocking current flow, as with a typical PN-junction diode, since there is no p-type semiconductor material and, consequently, no minority carriers (holes). As a result, a Schottky diode responds to bias changes very quickly and exhibits the traits of a rectifying diode.

Fig. 2.26 depicts the optimum band diagram of a certain metal and semiconductor (n-type) before contact. The reference level is the vacuum level. The parameter $\phi_s$ denotes the work function of semiconductor and $\chi$ is referred to as electron affinity. The parameter $\phi_m$ denotes the work function of metal (measured in volts). The fermi level is very close to the conduction band because the semiconductor is N-type material.

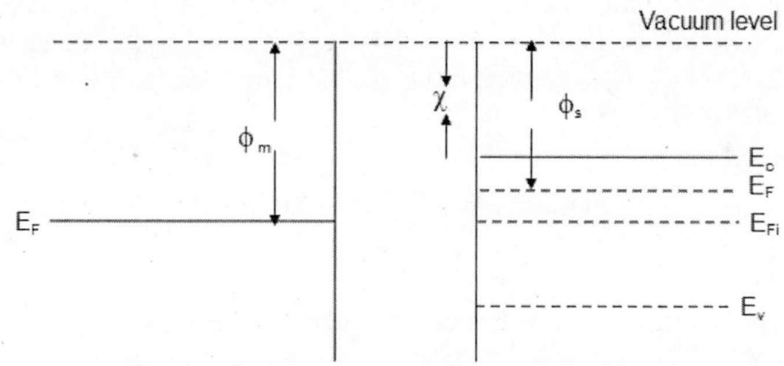


**Fig. 2.26:** Band Diagram of Schottky Diode

## 2.7.2 I-V Characteristics of Schottky Diode

In both biasing conditions, a Schottky diode's junction barrier is low compared to typical p-n junction diode. Schottky diode has a 0.25V as a barrier potential as opposed to a 0.7 V barrier potential for regular diodes. Fig. 2.27 compares the properties of a Schottky diode with a typical p-n junction diode.

Depending on the metal electrode employed, the forward current of a silicon Schottky diode can be significantly higher than that of a standard PN-junction diode. A smaller forward voltage drops for a given diode current, ID will result in reduced forward power dissipation in the form of heat across the junction. Since Ohm's law states that power equals to the product of volts and amps (P = V*I).

The Schottky diode is an excellent option in high current and low-voltage applications like solar photovoltaic panels because it has a smaller power loss than a normal pn-junction diode, which will result in reduced thermal dissipation.

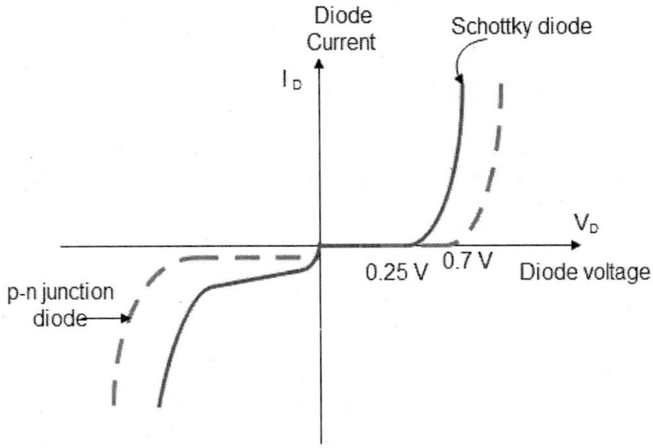


**Fig. 2.27:** I-V Characteristics of Schottky Diode

However, it should be noted that a Schottky diode's reverse leakage current ($I_R$) is often significantly higher than that of a pn-junction diode. But keep in mind that an Ohmic contact is present if the I-V characteristics curve displays a more linear non-rectifying feature. Ohmic contacts are frequently used to link semiconductor wafers and chips to the system's circuitry or external connection pins. For instance, attaching the pins of a Dual-In-Line (DIL) plastic packaging to the semiconductor wafer of a standard logic gate.

## 2.8 LIGHT EMITTING DIODE (LED)

A light emitting diode is a diode that emits light when it is forward biased. It relies on electroluminescence to function. In contrast to the photovoltaic effect, electroluminescence is a process that converts an electrical input into a light output. When current passes through a light-emitting diode (LED), electrons in the semiconductor mix with electron holes again to release energy in the form of photons.

### 2.8.1 Construction and Symbol

As seen in Fig. 2.28 a three semiconductor layers are employed on the substrate for LED. There is an area termed the active region located among the p-type and n-type regions. This area is in charge of the light's emission. All the way around the multilayer construction, the LED emits light. In order for the light to reflect in the correct exit direction, this layered structure is set within a small reflecting cup. Fig. 2.28 (b) depicts the cup-shaped structure. The LED symbol is the same as the normal diode symbol, with the addition of two tiny arrows signifying light output. The Fig. 2.28 (c) shows the symbol diagram of LED.

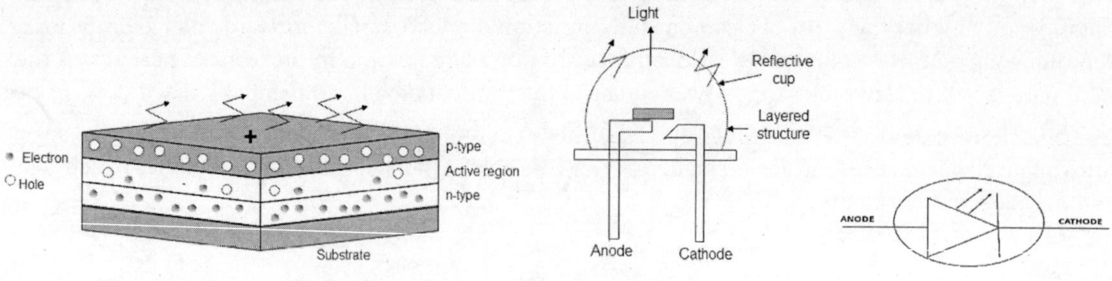


(a) (b) (c)

**Fig. 2.28:** LED: (a) Construction, (b) Cup Type Structure, (c) Symbol

## 2.8.2 Working Principle of LED

The LED works on the principle of electroluminescence. Minority electrons and holes are delivered from p to n and n to p, respectively, when the diode is forward biased. Minority carriers are more concentrated around the junction boundary. The majority charges carriers depicted in Fig. 2.29 recombine with the excess minority carriers at the junction.

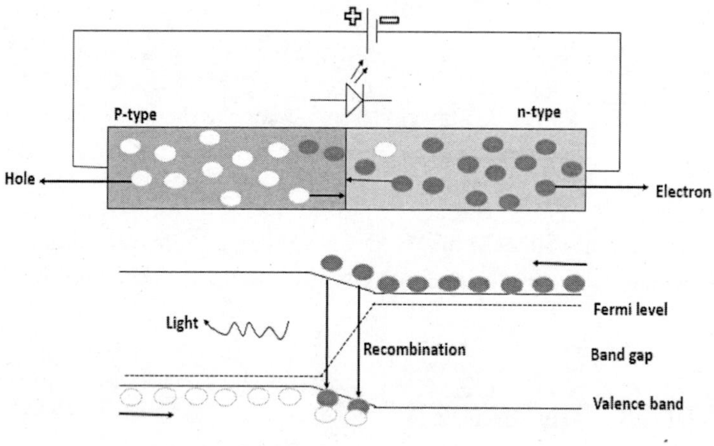


**Fig. 2.29:** Principle of operation of LED

The energy is released as photons after recombination. Conventional diodes emit the energy as heat. This phenomenon is known as Electroluminescence (EL). When an electric current is passed through a substance, it emits light, which is an electrical and optical phenomenon known as electroluminescence. As the forward voltage increases, the light intensity increases and reaches its maximum.

## 2.8.3 Materials Used for LED

The colour of an LED is influenced by the material used in the semiconducting element. The two major materials used in LEDs are indium gallium nitride alloys and aluminium gallium indium phosphide alloys. While red, orange, and yellow light is produced by aluminium alloys, green, blue, and white light is produced by indium alloys. Depending on the composition of these alloys, there are subtle variations in the colour of the light that is emitted. Table 2.2 lists the various materials and colours that were obtained.

**Table 2.2:** Different kind of Materials

| *Component* | **Wavelength** | **Colour** |
|---|---|---|
| GaP | 565 | Green |
| GaAsP | 590 | Yellow |
| GaAsP | 632 | Orange |
| GaAsP | 649 | Red |
| GaAlAs | 850 | Near IR |
| GaAs | 940 | Near IR |

### 2.8.4 Different Types of LED

- Alphanumeric LED
- Red Green Blue LEDs
- High-Power LEDs
- Flash LED
- Bi and Tri-Colour LEDs
- Lighting LED

## 2.9 PHOTODIODE

A photodiode is a specific kind of PN junction semiconducting device. This specific type of light sensor transforms light into electrical energy (voltage or current). There is an intrinsic layer that exists between the p (positive) and n (negative) layers. The photo diode turns light energy into electric current by absorbing it as an input.

Other names for it are photodetector, photo sensor, and light detector. The negative electrode of the battery (or power supply) is connected to the photodiode's p-side, while the positive electrode is connected to the photodiode's n-side. Common photodiode components include silicon, indium gallium arsenide phosphide, germanium, and indium gallium arsenide.

### 2.9.1 Symbol

It has two terminals, as shown in Fig. 2.30. The shorter and longer terminals are the cathode and anode, respectively. The circuit symbol of a photodiode is similar to that of an LED (Light-Emitting Diode), with the main difference being the direction of the arrows.

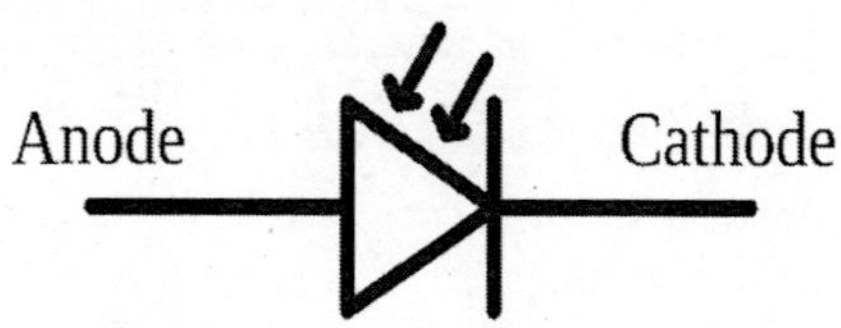


**Fig. 2.30:** Symbol of Photodiode

In an LED symbol, the arrows typically point outward, indicating the emission of light. However, in a photodiode symbol, the arrows point inward or towards the photodiode, representing the detection or absorption of incoming light.

### 2.9.2 Working Principle of Photodiode

When a light is used to illuminate the PN junction, covalent bonds are ionised. Thus, electron and hole pairs are created. Photocurrents are produced when electron-hole pairs form. Electron-hole pairs form when photons with energy greater than 1.1eV clash with the diode. A photon has a large energy impact on the atom when it enters the depletion zone of a diode. This results in the liberation of an electron from the atom's structure. After the electron is expelled, free electrons and holes are created.

An electron has a negative charge, whereas a hole carries a positive charge. The depletion of energy will incorporate an electric field. The electric field causes pairs of electron and hole to depart away from the junction. Electrons go to the cathode to create photocurrent, while holes move to the anode.

Photon energy and photon absorption intensity are closely connected. When photo energy is lower, absorption would be higher. This entire approach is known as the Inner- Photoelectric Effect.

Intrinsic and extrinsic excitations are the two types of photon excitation. Intrinsic excitation occurs when a photon moves an electron from the valence band to the conduction band. Extrinsic excitation, involves the influence of impurities or dopants introduced intentionally into the semiconductor material.

### 2.9.3 Operating Modes of Photodiode

#### *Photovoltaic Mode*

Zero bias mode is yet another name for this. When a photodiode is employed in low frequency and ultra-low light applications, this mode is preferable. A photodiode generates voltage when it is subjected to a flash of light. The voltage produced will be non-linear and have a limited dynamic range. When the photodiode is set up with OP-AMP in this mode, there will be a very minimal temperature change.

#### *Photoconductive Mode*

In this operating mode, the photodiode will function in negative biased mode. As the reverse voltage rises, the depletion layer's width grows. As a result, there will be a decrease in reaction time and junction capacitance. This method of functioning moves rather quickly and makes noise from electricity. Transimpedance amplifiers are preamplifiers for photodiodes. The photo diode can operate in the photoconductive mode since the modes of such amplifiers maintain the voltage at a constant level.

#### *Avalanche Diode Mode*

In this operating condition, the photo diode works at a high reverse bias state. It enables the multiplication of each photogenerated electron-hole pair through an avalanche breakdown. The photodiode thus produces internal gain. The built-in gain enhances the device's responsiveness.

#### *Linking a Photodiode to an External Circuit*

A photodiode works in a reverse biased circuit. The cathode is connected to the circuit's positive supply voltage, while the anode is connected to the circuit's ground. When the light strikes the device, current moves from the cathode to the anode. When used with external circuits, photodiodes are coupled to a power source within the circuit. A photodiode will only generate a tiny amount of current. An electrical device cannot be powered by this current. The circuit receives more current when connected to an external power source. Thus, the battery functions as a power source. The battery supply helps to increase the current value, which improves the performance of external machinery.

Scan to know more

## 2.10 SOLAR CELL

The solar cell is nothing more than a photovoltaic cell that operates according to the photovoltaic effect. The solar cell produces voltage according to the amount of incident light it receives. As a result of the photovoltaic effect, a solid-state device known as a solar cell transforms light energy into external energy.

### 2.10.1 CONSTRUCTION AND WORKING

Fig. 2.31 (a) depicts how a solar cell is built, whereas Fig. 2.31 (b) depicts the symbol for a solar cell. Light can penetrate to the junction due to the extremely thin surface layer of p-type material. The positive output terminal is the nickel-plated ring surrounding the p-type. The negative output terminal is the nickel-plated ring that surrounds the n-type material.

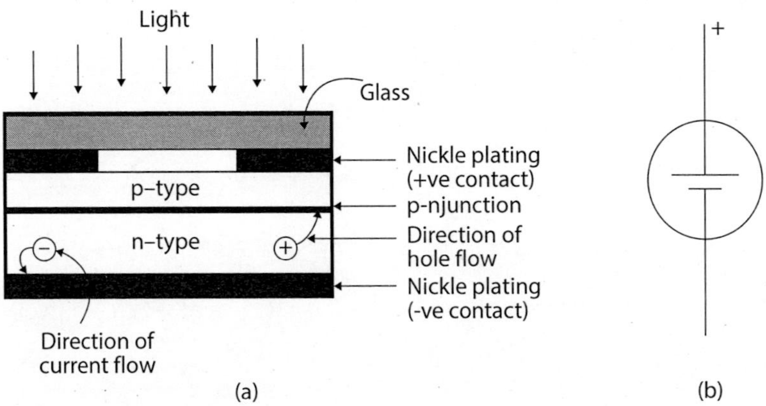


**Fig. 2.31:** Solar Cell: (a) Construction, (b) Symbol

When light enters the cell, the semiconductor material absorbs the photon's light energy and produces a number of electron-hole pairs. On both sides of the intersection, this occurs. Due to the presence of electric field, the electrons are directed towards the n-region whereas the holes are directed towards the p-region. This movement creates a minority current across the junction, which causes the voltage to form in the p and n regions. The most common ingredients for solar cells are silicon and selenium. The solar cells also use gallium arsenide, indium arsenide, and cadmium sulphide.

The efficiency of the solar cell is determined by the ratio of electrical power output to light source input. Usually, it ranges from 15 to 40 %.

$$\%\eta = \frac{P_O\,(\text{electrical})}{P_i\left(\text{light energy}\right)} \times 100 \qquad (2.30)$$

### 2.10.2 Characteristics of Solar Cell

A fraction of the light that enters the cell is absorbed by the semiconductor material. This indicates that the semiconductor has received the energy of the absorbed light. When the valence electrons have enough energy, they will break the bond with the parent atom and go free. Free electrons and holes are created as a result of this. Each side of the intersection will experience this phenomenon.

As was said for the basic p-n junction, the freshly created electrons in the p-type material are minority carriers and will travel quite readily across the junction. The discussion is the same for holes made in n-type material.

This causes a rise in the minority carrier flow, which has the opposite direction as a p-n junction's typical forward current. The properties of a solar cell are shown in Fig. 2.32. There are two limiting cases. First one is the short circuit condition that occurs at V=0. The current in this case known as short circuit current ($I_{sc}$).

$$I = I_{sc} = I_L \tag{2.31}$$

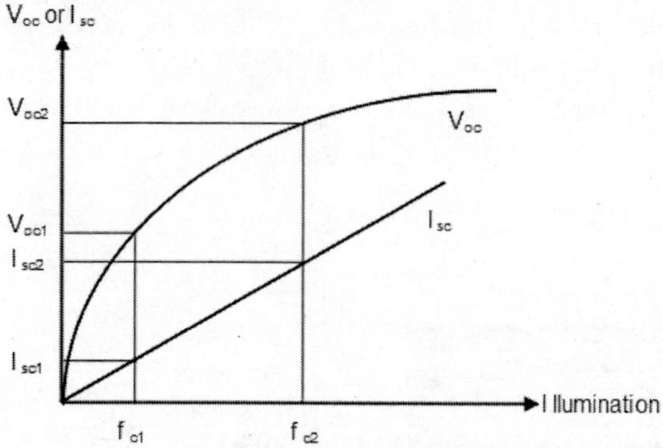


**Fig. 2.32:** Illumination of Solar Cell

The second limiting case is open-circuit condition. The net current is zero and voltage produced is open circuit voltage ($I_{sc}$).

$$I = 0 = I_L = I_S \left[ \exp\left(\frac{eV_{oc}}{KT}\right) - 1 \right] \tag{2.32}$$

We can find the open circuit voltage is,

$$V_{oc} = V_t \ In\left(1 + \frac{I_L}{I_S}\right) \left(\because V_t = kT/e\right) \tag{2.33}$$

Scan to know more

## UNIT SUMMARY

- In thermal equilibrium, the generation and recombination rates of electrons and holes must be equal in number.

$$G_{no} = G_{p0}$$

- Similarly, the recombination rate at thermal equilibrium is,

$$R_{no} = R_{p0}$$

- The excess electron and hole concentrations can be found using,

$$\delta n = n - n_o$$

$$\delta p = p - p_o$$

- Poisson's equation in electrostatics is given by,

$$\nabla.((\hat{\epsilon} \ \nabla \ \psi) = -\rho)$$

- The continuity equation is given by,

$$\nabla \cdot J + \frac{\partial \rho}{\partial t} = 0$$

- The barrier potential needs to be applied is given by,

$$V_{bi} = KT \ln (N_A \; N_D / n_i^2)$$

- Q Point: The intersecting point of diode response and load line.

$$Q_{point} = (I_D Q, V_{DQ})$$

- The impact of load resistance $R_L$ on the load line,

$$\text{Slope} = -\frac{1}{R_L}$$

The value of load resistance has a huge impact on the load line, because the load line changes with respect to the slope of the line.
- The small signal resistance of diode will be found using,

$$r_d \; \alpha \; \frac{1}{I_D}$$

- The efficiency of the solar cell is given by,

$$\%\eta = \frac{P_O\,(\text{electrical})}{P_i\left(\text{light energy}\right)} \times 100$$

- The width of the depletion region,

$$W = \sqrt{\frac{2\varepsilon}{q}\left(\frac{1}{N_d} + \frac{1}{N_a}\right) V_{bi}}$$

- The width of the depletion region on the p-side and on the n-side:

$$X_p = N_a/N_a + N_d \;\; W, \quad x_n = N_a/N_a + N_d \;\; W$$

- Junction profile:
In an abrupt junction,

$$W \; \alpha \; \sqrt{V_{bi}}$$

In a linearly graded junction,

$$W \; \alpha \; \sqrt[3]{V_{bi}}$$

It is possible to express the maximal electric field in the reverse bias as,

$$E_{max} = -\frac{2\left(V_{bi} + V_R\right)}{W}$$

## EXERCISES

### Multiple Choice Questions

**2.1 Consider a pn junction, where there is a difference in the doping on the both sides. Then the space charge region will be extended more into _________ region**

a. Lightly doped  b. Heavily doped
c. Middle of the junction  d. Can't determine

**2.2 Schottky-barrier diode is a**

a. Polar  b. Unipolar
c. Bipolar  d. Can't determine

**2.3 Schottky- barrier diode has**

a. High forward voltage drop  b. Low forward voltage drop
c. High reverse voltage drop  d. Low reverse voltage drop

**2.4 Schottky diodes are used as**

a. Biased clipper  b. Clamper
c. Clipper  d. Biased clamper

**2.5 The wavelength of the light in LED is depends on**

a. Holes  b. Electrons
c. Band gap  d. Diode

**2.6 The characteristics of LED diode similar to**

a. PN junction diode  b. Zener diode
c. Schottky diode  d. Photodiode

**2.7 Light generated in LED through**

a. Recombination of polarity charges  b. Generation of holes
c. Generation of electrons  d. Recombination of holes

**2.8 Solar cell is basically a**

a. PN junction diode  b. Zener diode
c. Schottky diode  d. Photovoltaic diode

**2.9 Materials which are used for solar cell must have band gap close to**

a. 1.5ev  b. 1ev
c. 0.5ev  d. 1eV

**2.10 Photo diode is constructed to function in**

a. Forward bias  b. Reverse bias
c. Both  d. None of the above

**2.11 Which of the following is not the drawback of Zener diode shunt regulator?**

a. The output voltage is fixed
b. The output voltage can vary with temperature

c. Variation in load current needs to be minimal

d. It is difficult to design

**2.12 Zener diode is used for, which application?**

a. Oscillator

b. Regulator

c. Rectifier

d. Filter

**2.13 The different term of bias point is**

a. Quiescent point

b. Node point

c. Terminal point

d. Static point

**2.14 The forward region of a semiconductor diode characteristic curve is where diode appears as**

a. High resistance

b. An ON switch

c. An OFF switchd.

A capacitor

**Answer of Multiple Choice Questions**

| | | | | |
|---|---|---|---|---|
| 2.1 **a** | 2.2 **b** | 2.3 **b** | 2.4 **d** | 2.5 **c** |
| 2.6 **a** | 2.7 **a** | 2.8 **d** | 2.9 **a** | 2.10 |
| **b** | 2.11 **d** | 2.12 **b** | 2.13 **a** | 2.14 **(b).** |

## SHORT AND LONG ANSWER TYPE QUESTIONS

### Category I

2.1 Define knee voltage.

2.2 What is meant by depletion region?

2.3 What is meant by Zener breakdown?

2.4 Define avalanche breakdown.

2.5 Draw I-V characteristics of PN junction diode.

2.6 Define barrier potential at the junction.

2.7 Compare Zener diode with conventional diode.

2.8 State material used in LED.

2.9 What is photoconductive cell?

### Category II

2.1 A solar cell is a pn junction device with no voltage directly applied across the junction. If it is so, how does a solar cell deliver power to load?

2.2 Explain the construction and working principle of LED.

2.3 Write a note on solar cell.

2.4 With a neat diagram explain the working of a PN junction diode in forward bias condition.

2.5 Explain the I-V characteristics of diode.

2.6 Explain the construction and working of Schottky diode.

2.7 Discuss the performance of Zener diode.

## Numerical Problems

2.1 Determine the built-in potential and depletion region width for the reverse biased pn junction. Take silicon pn junction at room temperature having doping concentration of , and the reverse bias of .
**(Ans: 0.695V, W= 5.262 $\times 10^{-4}$ cm)**

2.2 Draw the load line for the circuit shown in Fig 2.17. The circuit having the value of load resistance as 20Ω and the input voltage of 8V.
**(Ans: 0.4,0 and 8V,0)**

2.3 A diode has a voltage of 0.7V and a current of 100mA. What is the diode power?
**(Ans: 2.2W)**

2.4 In series are two diodes. The voltage of the first diode is 0.75 volts, while that of the second is 0.8 volts. What is the current flowing through the second diode if the first diode's current is 400mA?
**(Ans: 5mA)**

2.5 If the resistor is doubled, calculate the load current for the below diagram.
**(Ans:4.3mA)**

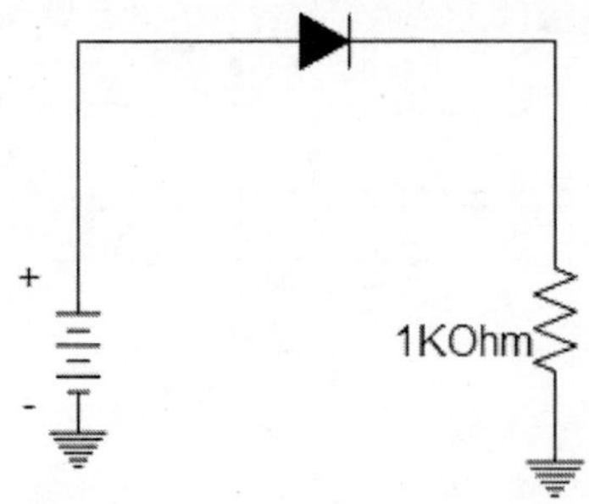


## PRACTICAL

## Diode Characteristics in A) Forward Bias B) Reverse Bias

### Aim

i. To study the p-n junction diode properties under forward and reverse bias conditions.
ii. To investigate the properties of Zener diodes under forward and reverse bias conditions.

### Apparatus Required

- PN Junction diode
- Zener diode
- Power supply (0-30)V
- Ammeter (0-30) mA
- Ammeter (0-100)
- Voltmeter- (0-1) V, (0- 30) V

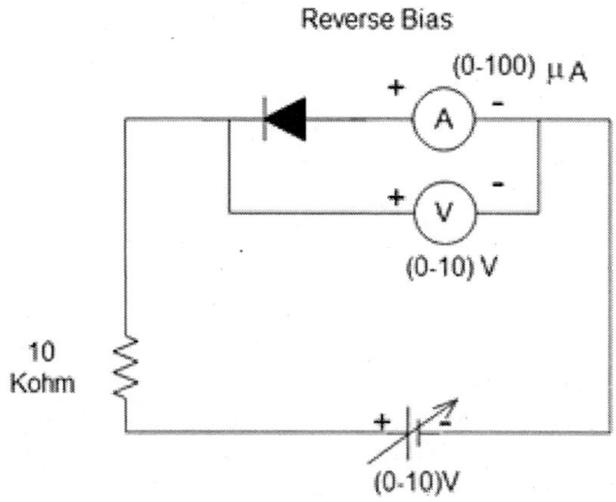


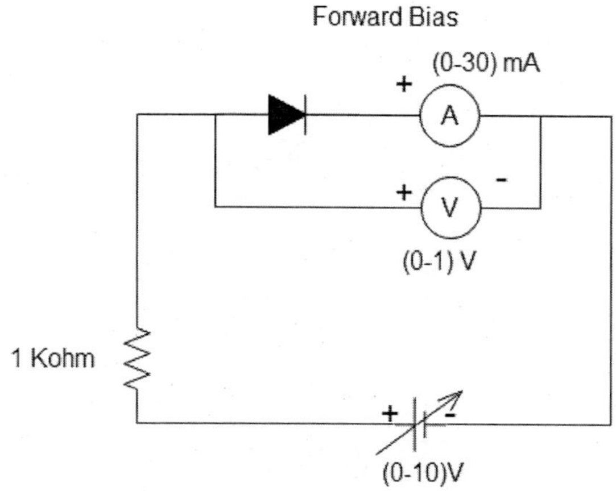


## Procedure

### *Forward Biasing*

1. Connect the circuit in accordance with the schematic.
2. Adjust the power supply so that readings are made on the voltmeter in steps of 0.1V until the power supply's needle indicates 20V.
3. Make a note of the associated ammeter reading.
4. Plot V vs I in the graph.
5. Find the dynamic resistance,
6. Replace PN Junction diode with Zener diode.
7. For Zener diode, repeat the same procedure with Zener diode.

### *Tabulation*

| S.No | Forward Voltage (v) | Current (mA) |
|---|---|---|
| | | |
| | | |
| | | |
| | | |

### *Reverse Biasing*

1. Connect the circuit in accordance with the schematic.
2. Adjust the power supply so that readings are made on the voltmeter in steps of 1V until the power supply's needle indicates 20V.
3. Make a note of the associated ammeter reading.
4. Plot V vs I in the graph.
5. Find the dynamic resistance, $r = \frac{\Delta V}{\Delta I}$
6. For Zener diode, repeat the same procedure with Zener diode.
7. Replace PN Junction diode with Zener diode.

## Tabulation

| S.No | Reverse Voltage (v) | Current (A) |
|---|---|---|
| | | |
| | | |
| | | |
| | | |

## Result

Forward and reverse bias characteristics of PN junction diode and Zener diode are plotted and their response is analyzed.

## KNOW MORE

In the initial stages of a device to use semiconductors as logic switches, the National Bureau of Standards Eastern Automatic Computer (SEAC) used 10,500 germanium diodes in 1950. They also performed functions as varied as high-voltage power supply rectifiers and memory storage components in later devices. Displays and input/output devices are still made possible by light emitting and sensing features for fiber-optic interconnections.

Scan to know more

### History

Frederick Guthrie made the discovery in 1873. A positively charged electroscope will be discharged by a red-hot metal ball supplied close to it but not a negatively charged one. The Edison effect, also known as a unidirectional current between heated and unheated sections of a bulb, was discovered by Thomas Edison in 1880. He was given a patent on the use of the phenomenon in a DC voltmeter. About 20 years later, John Ambrose Fleming, discovered that the Edison effect might be used as a radio detector by a former employee of Thomas Edison who is now a scientific advisor to the Marconi Company. The Fleming valve, the first actual thermionic diode in Britain, was granted a patent on November 16, 1904. (followed by U.S. Patent 803,684 in November 1905).

Vacuum tube electronics, such as radios, TVs, sound systems, and instruments, virtually always employed valve diodes at this time. Due to the development of selenium rectifier technology in the late 1940s and later semiconductor diodes in the 1960s,

**Frederick Guthrie**

gradually lost market share. They are still utilized today in a few high-power applications, where their superior durability and capacity to endure transient voltages provides them a competitive edge over semiconductor devices, as well as in applications for musical instruments and audiophile equipment.

## Application

A semiconductor diode's current-voltage characteristic can be tailored by selecting the semiconductor materials and the doping impurities added during production. These techniques can be used to create multifunctional special-purpose diodes. For example, diodes are used to produce radio-frequency oscillations (tunnel diodes, Gunn diodes, IMPATT diodes), control voltage (Zener diodes), protect circuits from high voltage surges (avalanche diodes), electronically tune radio and TV receivers (varactor diodes), safeguard circuits from high voltage surges (avalanche diodes), and produce light (lamp diodes) (light-emitting diodes). Negative resistance is shown by Tunnel, Gunn, and IMPATT diodes, which is advantageous in switching and microwave circuits. Diodes can be employed as shot-noise generators, both of the semiconductor and vacuum types.

## REFERENCES AND SUGGESTED READINGS


1. David.J.Griffiths, *Introduction to Quantum Mechanics,* Pearson Prentice Hall, 2005.
2. Chenming Calvin Hu, *Modern Semiconductor Devices for Integrated Circuits,* Pearson Prentice Hall, 2009.
3. S.O.KASAP, *Principle of Electronic Materials & Devices,* MC Graw Hill Education, 2018.
4. S.Datta, *Quantum Transport:Atom to Transistor,* Cambridge University Press, 2005.
5. Robert F. Pierret, *Semiconductor Device Fundamentals,* ADDISON-WESLEY Publishing Company, 1996.
6. Donald A. Neamen, *Semiconductor Physics and Devices Basic Principle,* MC Graw Hill Companies, 2012.
7. Ben G, Streetman, Sanjay Kumar Banerjee, *Solid State Electronic Devices,* Pearson Education Limited, 2016.
8. *https://physicsinmyview.com/2020/09/facts-about-quantum-physics.html*
9. *https://onlinecourses.nptel.ac.in/noc21_ee80/course*
10. *https://onlinecourses.nptel.ac.in/noc21_ee59/course*
11. *https://onlinecourses.nptel.ac.in/noc21_ee19/course*
12. https://en.wikipedia.org/wiki/Diode

# 03 APPLICATION OF DIODE

## UNIT SPECIFICS

*Through this unit we have discussed the following aspects:*

- *Understanding of diode circuit*
- *Use of a diode as a switch*
- *Diode as clipper and clamper circuits*
- *Application of Zener diode as a regulator circuit*
- *Application of photodiode*
- *Application of solar cell and LED*
- *Application of Schottky diodes*

*The themes' practical applications are covered in order to increase students' creativity and curiosity as well as their ability to solve problems.*

*Along with a large number of multiple-choice questions and questions with short and long answers divided into two categories based on Bloom's taxonomy's lower and higher orders, the unit also includes assignments through a number of numerical problems, a list of references, and suggested readings that one can use as practise materials. It is significant to notice that several portions of the website feature QR codes that may be scanned for further information on a variety of interesting topics.*

*Following the content-based relevant practical, there is a "Know More" section. This section has been thoughtfully created so that the additional material supplied will be useful to the book's readers. This section primarily highlights the initial activity, examples of some interesting facts, analogies, the history of the development of the subject focusing on the key observations and findings, timelines beginning with the development of the concerned topics up to the present, applications of the subject matter for our day-to-day real life or/and industrial applications on variety of aspects, case study related to environmental, sustainability, social, and ethical issues which applicable, and finally inquisitiveness and curiosity topics of the unit.*

## RATIONALE

*This unit on applications of diode helps students to get a primary idea about the application part of the diode. It explains diode as a switch and its operation. All these basic aspects are relevant to start the electronics properly. It then clearly explains Zener diode and its application as a regulator circuit. Further it explains about clipper and clamper circuit. Finally, the chapter ends with the application of optoelectronics device. Such as solar cell, Schottky diode, photo diode and LED. All these are discussed at length to develop the subject. Some related problems are pointed out with suitable diagrams which can help further for getting a clear idea of the concern topics on diodes.*

*The basic function of the diode to change the AC signal into DC signal. This is by removing a portion of the input signal. We can use a diode in rectifier circuit. The light emitting diodes are used in illumination devices. Regulator applications are also discussed in this chapter.*

## PRE-REQUISITES

*Fundamentals of Semiconductor Diodes*

## *UNIT OUTCOMES*

*List of outcomes of this unit is as follows:*

*U3-O1: Describe diodes in a switching circuit*

*U3-O2: Describe Zener diode as voltage regulator*

*U3-O3: Explain the operation of clipper and clamper circuit*

*U3-O4: Realize the application of Schottky diode*

*U3-O5: Application of solar cell, photo diode, LED*

| Unit-3 Outcomes | EXPECTED MAPPING WITH COURSE OUTCOMES (1-Weak Correlation;2-Medium Correlation;3-Strong Correlation) | | | | | |
|---|---|---|---|---|---|---|
| | CO-1 | CO-2 | CO-3 | CO-4 | CO-5 | CO-6 |
| U3-01 | 3 | 3 | 3 | - | 3 | 1 |
| U3-02 | 1 | 1 | 2 | 2 | 1 | - |
| U3-03 | 2 | 1 | 3 | 1 | 2 | 1 |
| U3-04 | - | - | 3 | 1 | 2 | 2 |
| U3-05 | 3 | 3 | 3 | - | 3 | 1 |

## 3.1 DIODE AS A SWITCH

The interface or intersection of semiconductor material p & n types within a semiconductor is known as a P-N junction. Have a closer look at the diode symbol shown in Fig. 3.1. It resembles an arrow, indicating the current's direction and permits anode to cathode current to flow. It is a voltage-controlled, two-terminal device. It possesses switch-like qualities. It permits current to flow in one direction but does not allow it in the opposite direction.

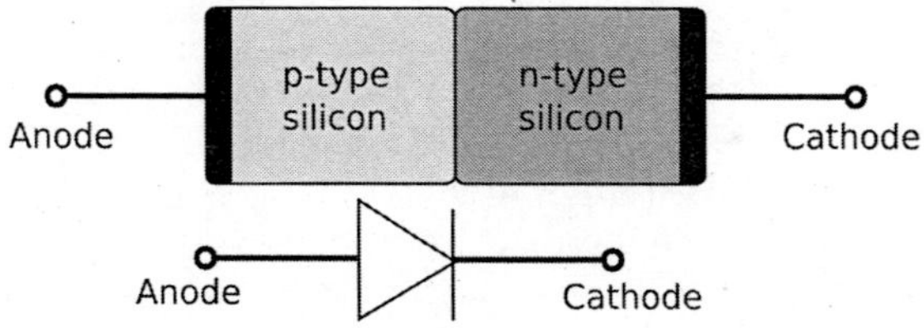


**Fig. 3.1:** Representation of Diode Symbol

In other words, a diode should have negligible resistance under forward biased circumstances. It functions like a closed switch. When reverse biased, it operates as an open switch. Signal diodes are another name for diodes used in switching applications.

### 3.1.1 Diode Switching Circuit

A diode can be used as a switch in electronic circuits. When used in this manner, a diode is typically operated in either forward-biased or reverse-biased state to control the flow of current.

**Forward-Biased Diode as a Switch:** When a diode is forward-biased, the positive terminal of the voltage source is connected to the anode (P-side) and the negative terminal to the cathode (N-side), so that the diode allows current to flow through it.

**Reverse-Biased Diode as a Switch:** When a diode is reverse-biased, meaning the positive terminal of the voltage source is connected to the cathode (N-side) and the negative terminal to the anode (P-side), so that the diode blocks the flow of current.

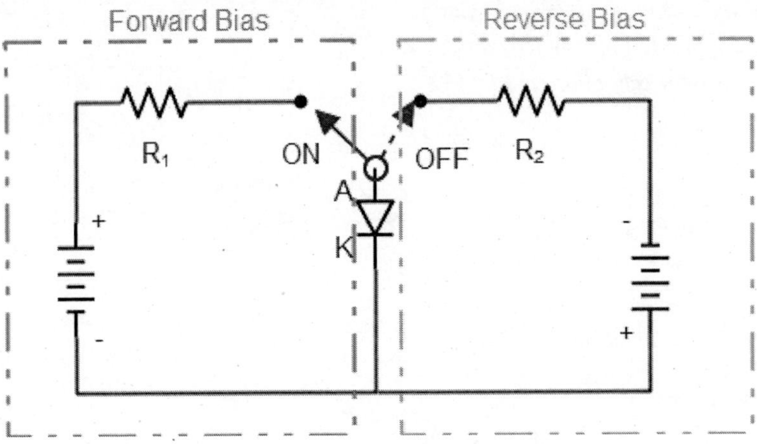


**Fig. 3.2:** Switching Circuit using Diode

The best circuit to describe how a diode switches in a circuit is shown in Fig. 3.2, where the diode is connected to biasing networks for forward and reverse bias. Since the forward biasing network makes up the circuit's first half, it will function as a closed circuit at that point.

The reverse biasing network makes up the circuit's second half, and at that point, the circuit will function as an open circuit. This is how the diode function as a switch when biasing switches from forward to reverse and reverse to forward.

### 3.1.2 Working

1. Apply a positive voltage difference between the anode and cathode. The diode acts as a closed switch when the anode is more positive than the cathode. The input and output are directly connected. Because of this, the current moves from positive to negative terminals.
2. Now change the polarity. Consequently, the cathode has a higher potential than the anode. The diode functions as an open switch between input and output terminals in this scenario. As a result, the diode has no current flowing through it.

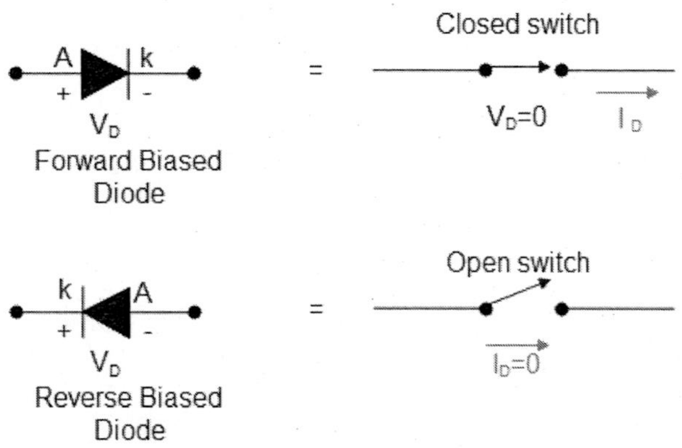


**Fig. 3.3:** Operation of Diode under Forward and Reverse Bias

## 3.2 CLAMPING AND CLIPPING CIRCUIT

In the previous section, we saw the diode as a switching circuit. Apart from this, a diode has many other applications. Here we are going to see about the most useful applications of diodes clippers and clampers. We use the diodes along with *resistors and capacitors* to shape the waveforms. *Clippers* can be used to *clip* a section of the input waveform. *Clampers* can be used to *move or clamp* the dc voltage level.

### 3.2.1 Clippers

In networks called clippers, the input signal is clipped without changing the rest of the waveform in any way. Two different categories for clippers. They are parallel and series clippers. In a series clipper, the load resistance is linked with the clipper diode by a series connection. The diode and resistance with load are linked in parallel for parallel clippers. Again, there are two types of series clippers. Unbiased series clippers are the first ones. Biased series clippers are the second one. Parallel clippers are also classified into two types: unbiased parallel clippers and biased parallel clippers.

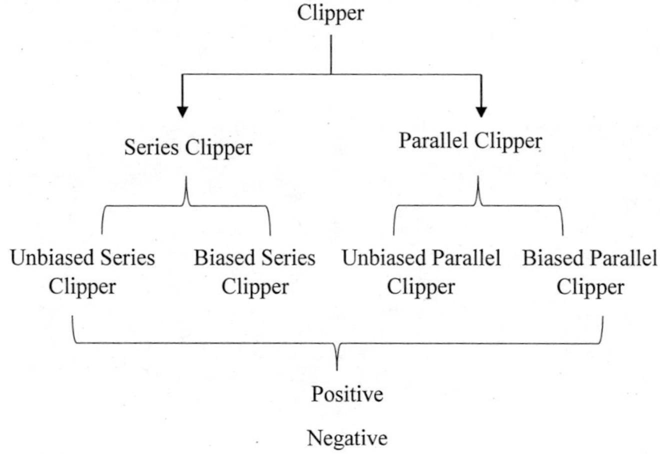


**Fig. 3.4:** Classification of Clippers Circuit

## *Unbiased Series Positive Clipper*

To analyze this kind of clipper circuit, let us, consider the circuit which has a diode series-connected to a Load Resistance ($R_L$). Fig. 3.5 displays a circuit schematic for an unbiased series positive clipper. The arrowhead of the diode is connected to the input. The function of the series clipper is to clip off the positive portion of the input waveform.

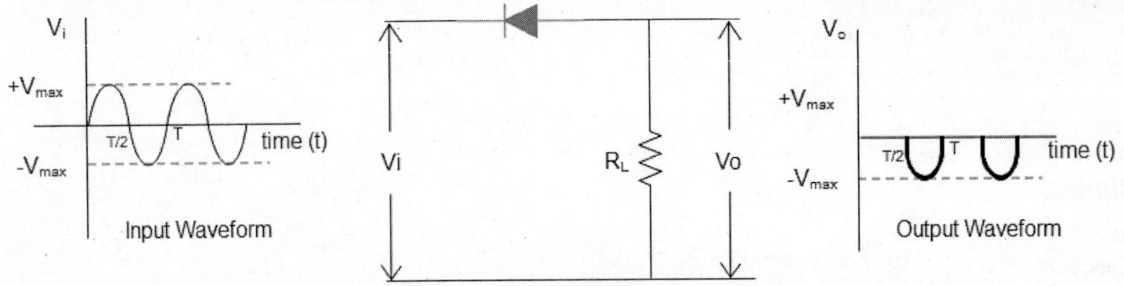


**Fig. 3.5:** Unbiased Series Positive Clipper

In the positive half cycle, the cathode of the diode is connected to the input's positive terminal, while the negative terminal is connected to the anode. Now, the bias is reversed, and the diode functions as an open switch. So, $V_0$ is

$$V_0 = 0, \quad 0 \leq t \leq T/2$$

The cathode is connected to the negative terminal of the input during the negative half cycle, while the anode is connected to the positive terminal. Currently, the diode operates in a forward bias situation and functions as a closed switch. So, the output becomes,

$$V_0 = V_i, \quad T/2 \leq t \leq T$$

### Unbiased Series Negative Clipper

To study the unbiased series negative clipper circuit, which is depicted in Fig. 3.6, the diode anode terminal is connected to the input waveform. The main goal of this unbiased series negative clipper is to remove the negative portion of the input waveform. In all these analysis diode characteristics is assumed as ideal and hence the drop across the diode is treated as 0V when it is in forward bias.

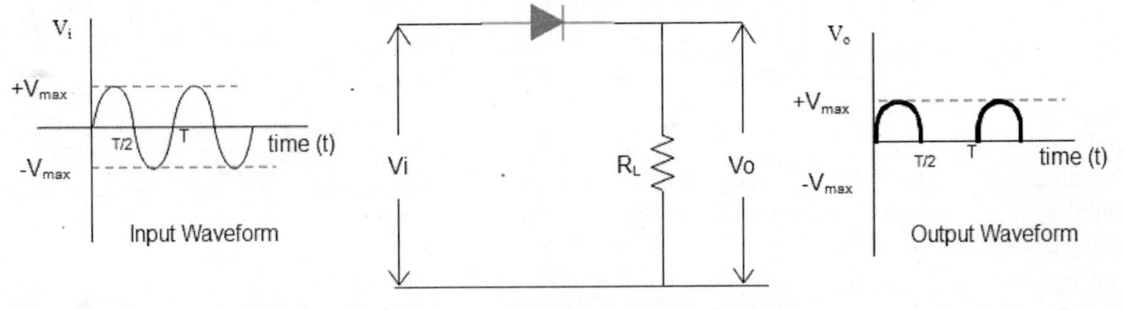


**Fig. 3.6:** Unbiased Series Negative Clipper

The anode and cathode are connected to the input's positive and negative terminals, respectively, when the input is positive half cycle. At this point, the diode works under a forward bias mechanism and performs like a connected switch. Then the output is,

$$V_0 = V_i, \quad 0 \leq t \leq T/2$$

The negative terminal is connected to the anode during the negative half cycle and the positive terminal to the cathode. The circuit works under reverse bias conditions and acts like an open switch, the output becomes,

$$V_0 = 0, \quad T/2 \leq t \leq T$$

## *Biased Series Positive Clipper*

Biased clippers are used when there is a need to remove a small portion of the input signal. In some cases, it is preferred to eliminate the small amount from the given input signal. In such situation, biased clippers are used. This biased series positive clipper is classified into two types. One is positive biased series positive clipper; the other one is a negative biased series positive clipper.

i. **Positive Biased Series Positive Clipper**

The design is very similar to the series positive clipper. The major difference is that an additional positive biasing is added to the load. The below Fig. 3.7 shows the series positive clipper with positive biasing.

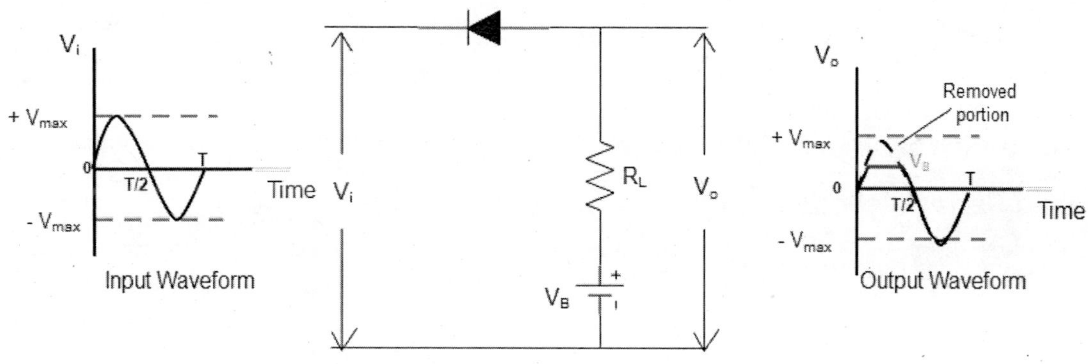


**Fig. 3.7:** Series Positive Clipper with Positive Biasing

In the positive half cycle, the diode operates at reverse bias by the input signal. But we apply another supply voltage of $V_B$ as shown in Fig. 3.7. Have a closer look, the cathode of the diode is linked to the supply's negative terminal, and its anode is connected to its positive terminal. The increased bias voltage of VB allows them to function in forward bias as a result. There are two cases we need to consider; one is when the input signal ($V_i$) is less than the bias voltage ($V_B$). In this case, the bias voltage $V_B$) dominates, making the diode operates in forward bias, so the input signal appears at the output. In another case, the input signal ($V_i$) is greater than the bias voltage ($V_B$), making the diode operates in reverse bias, so the signal appears at the output by an amount of $V_B$.

$$\left.\begin{array}{l} V_0 = V_i \; ; V_i < V_B \\ V_0 = V_i \; ; V_i < V_B \end{array}\right\} 0 \leq t \leq T/2$$

In the negative half cycle, the diode operates at forward bias by both input signal and bias voltage. So, the signal that appeared at the output is,

$$V_0 = V_{in}, \qquad T/2 \leq t \leq T$$

ii. **Negative Biased Series Positive Clipper**

Negative biasing is also very similar to positive biasing. The only difference is the polarity of the applied battery voltage. The following Fig. 3.8 shows negative biasing.

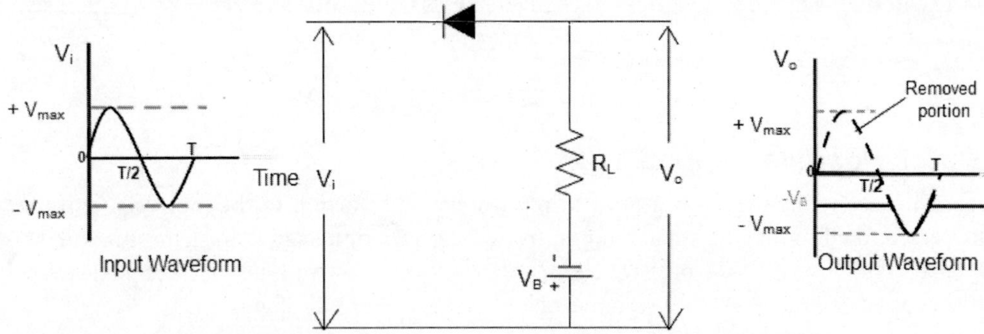


**Fig. 3.8:** Series Positive Clipper with Negative Biasing

As long as the input voltage is greater than or equal to $-V_B$ during the positive half cycle, the diode operates at reverse bias. Thus, the result shows battery voltage. As the battery voltage is negative, the same voltage with a constant amplitude is appear. As a result, the signal above $-V_B$ is completely clipped off.

During the negative half cycle, we need to consider two cases. The first case is, the input voltage ($V_i$) is greater than the battery voltage ($-V_B$), making the diode operate in reverse bias, so the result appears the same as the battery voltage. The second case is when, the input voltage ($V_i$) is less than the battery voltage ($-V_B$), making the diode operates in forward bias, so the input signal appears at the output.

## *Biased Series Negative Clipper*

The biased series negative clipper is classified into two types. One is a positive biased series negative clipper; the other one is a negative biased series negative clipper.

i. **Positive Biased Series Negative Clipper**

The design is very similar to the series negative clipper. The major difference is that an additional positive biasing is added to the load. The below Fig. 3.9 shows the series negative clipper with positive biasing.

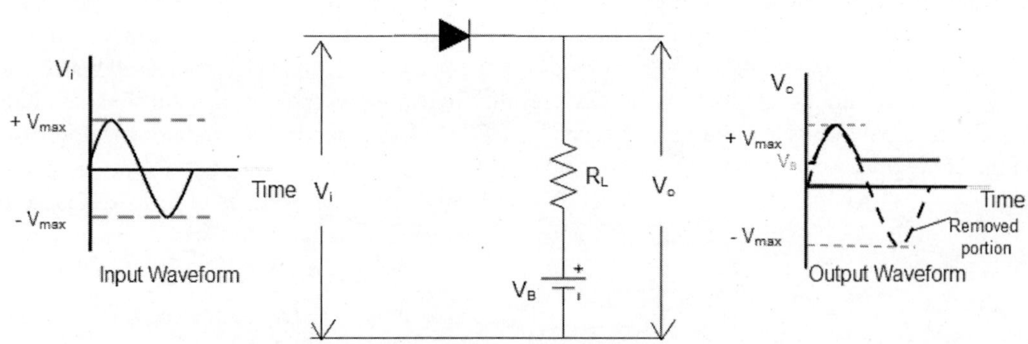


**Fig. 3.9:** Series Negative Clipper with Positive Biasing

In the positive portion of the cycle, if the input signal is less than the battery voltage, the diode is reverse biased and the signal appears at the output by an amount of $V_B$. As long as the signal exceeds the battery voltage, the diode is forward biased during the positive portion of the cycle. The input signal, therefore, emerges at the output.

The diode works at reverse bias during the negative half cycle. Because in the negative portion of the cycle, the input signal is always less than the battery voltage. As a result, the signal appears at the output by an amount of $-V_B$.

ii. **Negative Biased Series Negative Clipper**

Negative biasing is also very similar to positive biasing. The only difference is the polarity of the applied bias voltage. The following Fig. 3.10 shows a negative bias series negative clipper.

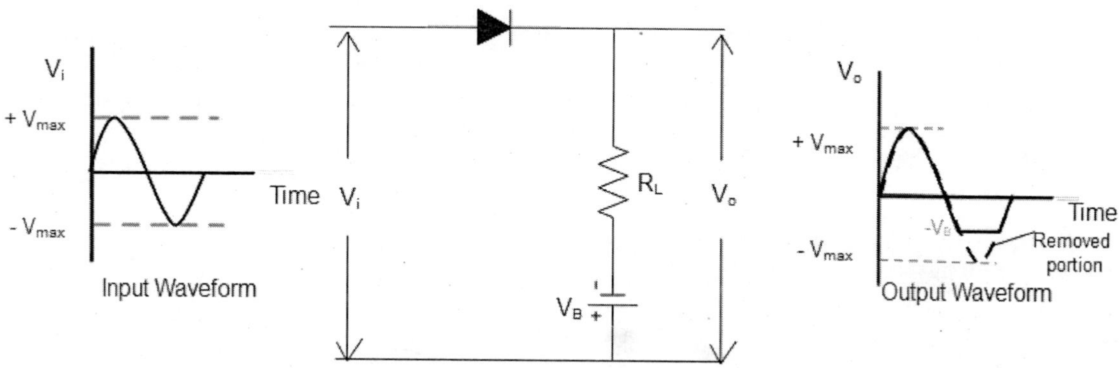


**Fig. 3.10:** Series Negative Clipper with Negative Biasing

The diode is forward biased by the input signal and battery voltage during the positive half cycle. The diode always remains forward biased, regardless of whether the input signal is larger or lower than the battery voltage.

During the negative half cycle, there are two cases in consideration. In the first case, the input voltage ($V_i$) is greater than the battery voltage ($-V_B$), making the diode operates at forward bias. So, the input signal appears at the output. The second case is, the input signal ($V_i$) is less than the battery voltage ($-V_B$), making the diode operates at reverse bias. So, the result shows battery voltage as the output.

## *Unbiased Parallel Positive Clipper*

To analyze this kind of clipper circuit, let us, consider the circuit that has a diode connected in parallel to the output voltage. The circuit diagram for the unbiased parallel positive clipper is shown in Fig. 3.11. The function of a parallel positive clipper is to clip off the positive portion of the input waveform.

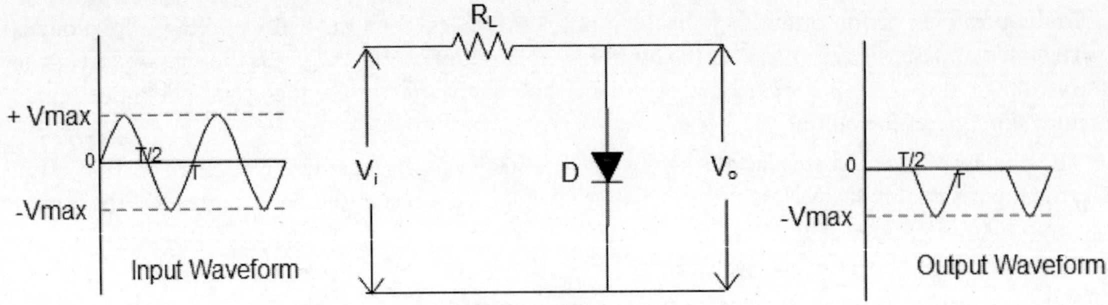


**Fig. 3.11:** Unbiased Parallel Positive Clipper

The parallel positive clipper passes the output signal during the reverse bias condition and does not allow the signal during the forward bias because of the parallel combination shown in Fig. 3.11.

The diode is forward biased during the positive half of the input cycle in parallel positive clipping. As a result, the output is 0.

$$V_0 = 0, \quad 0 \leq t \leq T/2$$

During the input's negative half cycle, the diode is biased in opposite direction. The result is the fraction of the input signal. The output $V_0$ becomes,

$$V_0 = V_i, \quad T/2 \leq t \leq T$$

## *Unbiased Parallel Negative Clipper*

Fig. 3.12 depicts a parallel negative clipper circuit; the operation is somewhat similar to the parallel positive clipper but it clips off the negative portion of the input signal.

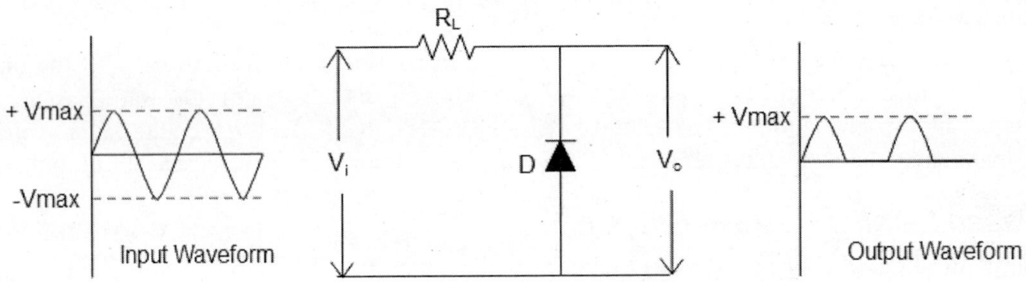


**Fig. 3.12:** Unbiased Parallel Negative Clipper

The diode operates in reverse bias when the input cycle is positive. So, whatever the input is given will appear in the output. When the input cycle is negative, the diode is in a forward bias condition. The output is zero.

## Biased Parallel Positive Clipper

This biased parallel positive clipper is classified into two types. One is a positive biased parallel positive clipper; the other is a negative biased parallel positive clipper.

i. **Positive Biased Parallel Positive Clipper**

The design is very similar to the parallel positive clipper. The major difference is that an additional positive bias is added to the diode. The below Fig. 3.13 shows the parallel positive clipper with positive biasing.

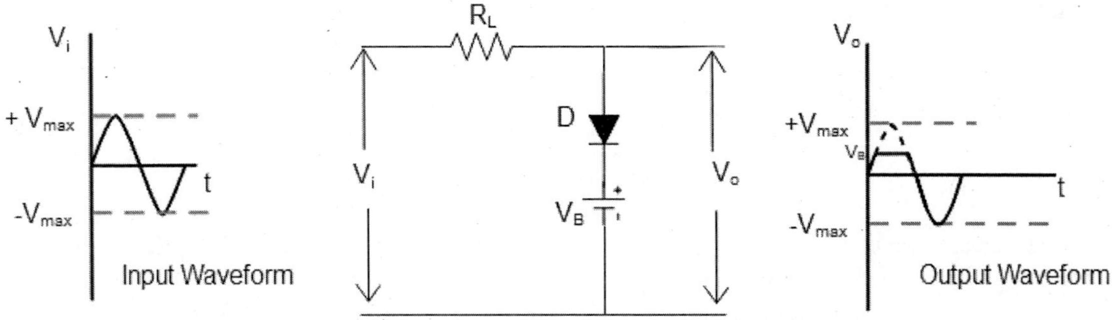


**Fig. 3.13:** Parallel Positive Clipper with Positive Biasing

In the Fig. 3.13, the cathode of the diode is attached to the positive terminal of the battery. Due to this, the diode is always reverse biased with the exception of situations when the input signal is more positive than the battery voltage (i.e., when $V_i > V_B$). If the input signal is more positive than the battery voltage, the diode is forward bias. So, the output is equal to the battery voltage. Remaining cases, the output is equal to the input.

ii. **Negative Biased Parallel Positive Clipper**

Knowing what happens if the battery voltage's polarity is reversed is interesting. Fig. 3.14 demonstrates negative biasing. Here, because the input is higher than the voltage -VB, the input signal is clipped off. The diode is forward biased by both supply voltages during the positive cycle. As a result, the output shows the battery voltage ($V_B$).

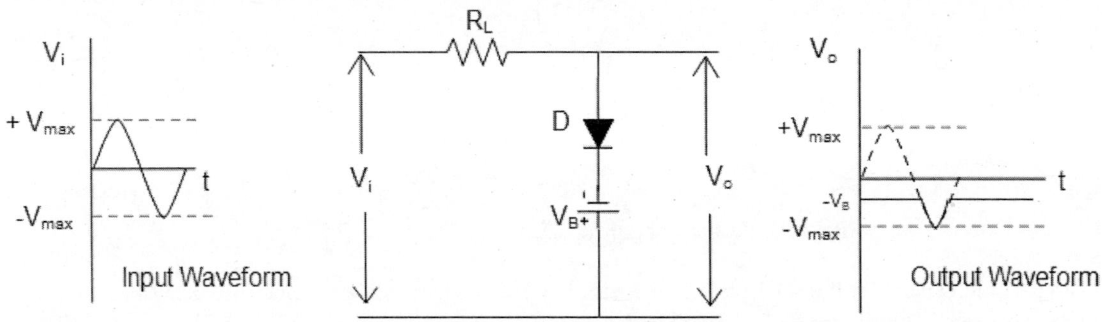


**Fig. 3.14:** Parallel Positive Clipper with Negative Biasing

During the negative half cycle, the input voltage Vi is initially lower than the battery voltage VB. The battery voltage has caused the diode to be forward biased as a result. So, the signal appears at the output by an amount of $V_B$. When the input signal exceeds $V_B$, the diode is biased in the opposite polarity. So, in this case, the output is showing the input signal.

## Biased Parallel Negative Clipper

Clipping occurs in this circuit during the negative half cycle. One is a positive biased parallel negative clipper and other is a negative biased parallel negative clipper.

i. **Positive Biased Parallel Negative Clipper**

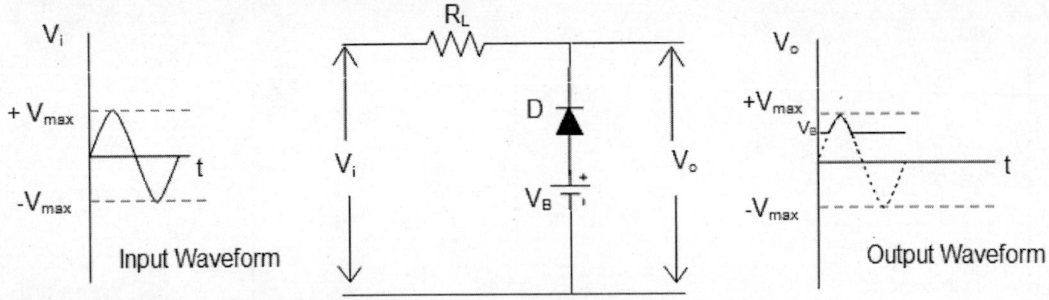


**Fig. 3.15:** Parallel Negative Clipper with Positive Biasing

There are two cases during the positive cycle. First, the input voltage is less than battery voltage (VB). So, the diode becomes forward bias and acts as a closed switch. As a result, the output displays the battery voltage, and signals that are below the battery voltage are clipped. Second, the diode becomes reverse biased and acts as an open switch. As a result, the output displays the input voltage, which is higher than the battery voltage.

The output shows the battery voltage ($V_B$) because the diode gets forward biased and the loop closes during the negative half cycle.

ii. **Negative Biased Parallel Negative Clipper**

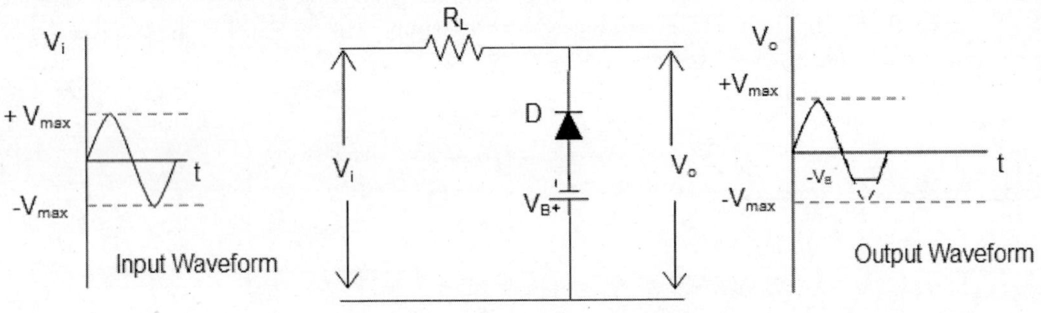


**Fig. 3.16:** Parallel Negative Clipper with Negative Biasing

When the cycle is in the positive half, the diode operates with reversed bias and serves as an open switch. As a result, the output displays the whole input voltage. A forward bias develops in the diode during the negative cycle, when the input is lesser than the battery voltage ($V_B$). The remaining signal is cut off in the negative cycle, up to the battery voltage.

## Combinational Clipper

In some applications there is a need to clip the input signal from both the positive as well as negative sides, this particular clipper makes use of parallel biased clippers. The below Fig. 3.17 shows the combinational clipper.

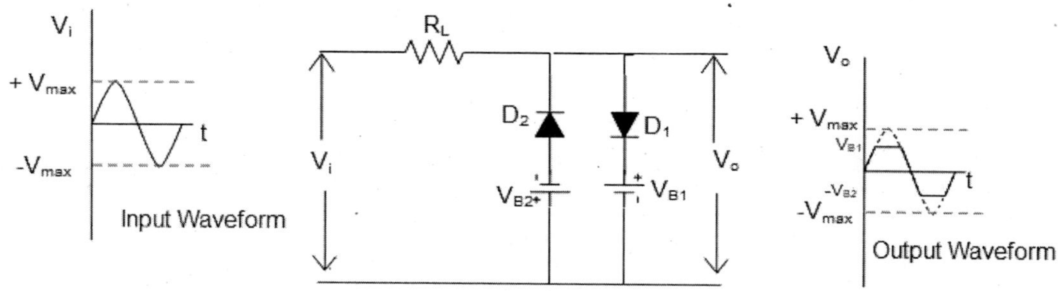


**Fig. 3.17:** Combinational Clipper

The input signal $V_i$ and the bias voltage $V_{B1}$ are used to forward and reverse bias the diode $D_1$ throughout the upward half cycle. On the other hand, the bias voltage $V_{B2}$ and input signal $V_i$ both works together to bias the diode $D_2$.

When the diode $D_1$ conducts during the positive cycle of the input signal, the bias voltage $V_{B1}$ manifests at the output. When the diode $D_2$ conducts throughout the negative half of the input signal, the bias voltage $V_{B2}$ manifests at the output. Therefore, to clip the output throughout both cycles, both diodes conduct alternately.

**Example 3.1:** Determine $V_0$ for the input shown.

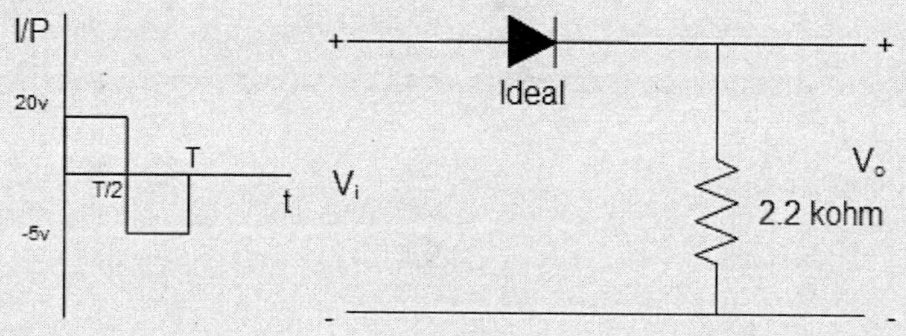


**Solution:**

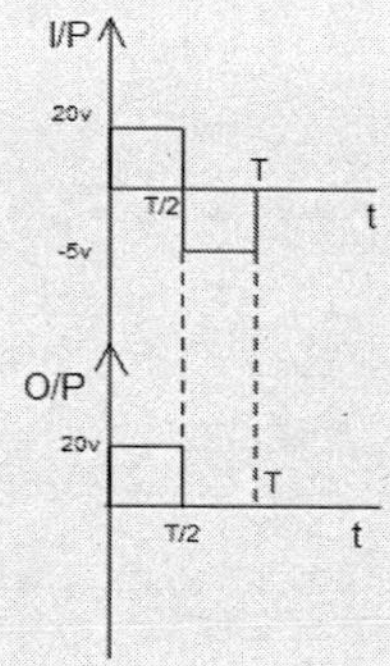


$$V_0 = V_i, \quad 0 \le t \le \frac{T}{2}$$

$$V_0 = 0, \quad T/2 \le t \le T$$

**Example 3.2:** Determine $V_0$ for the diagram given below.

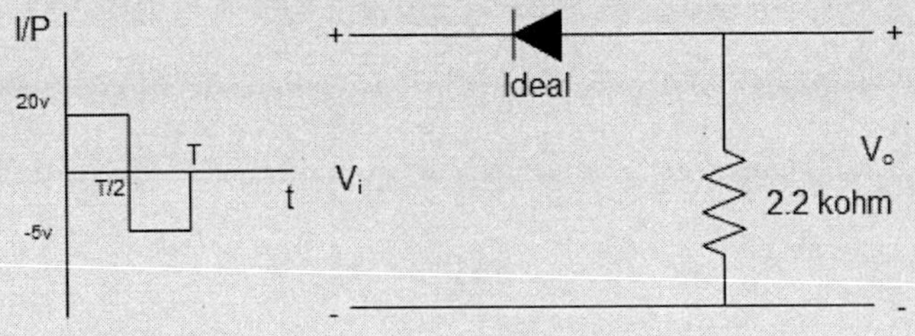


**Solution:**

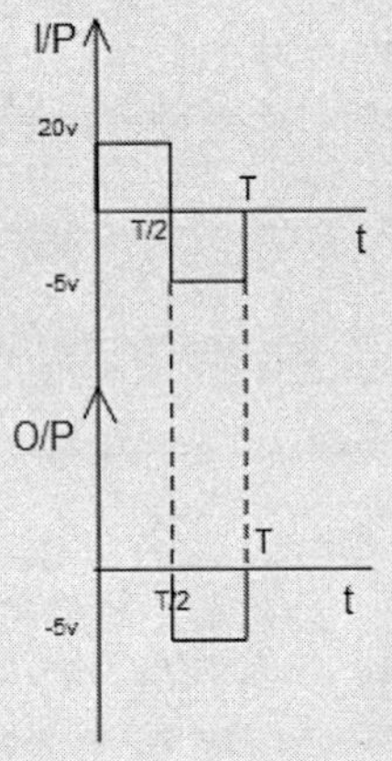


$$V_0 = 0, \quad 0 \leq t \leq \frac{T}{2}$$

$$V_0 = 5V, \quad T/2 \leq t \leq T$$

## 3.2.2 Clamper

A clamper is a network made of a diode, a resistor, and a capacitor that changes the applied signal's appearance while shifting the waveform to a new DC level. The circuit is classified into two types. Biased clamper and unbiased clamper. The functioning of the clamper circuit depends on capacitor's constant time. Constant time is denoted by the symbol $\tau$ and it is expressed as $\tau = RC$.

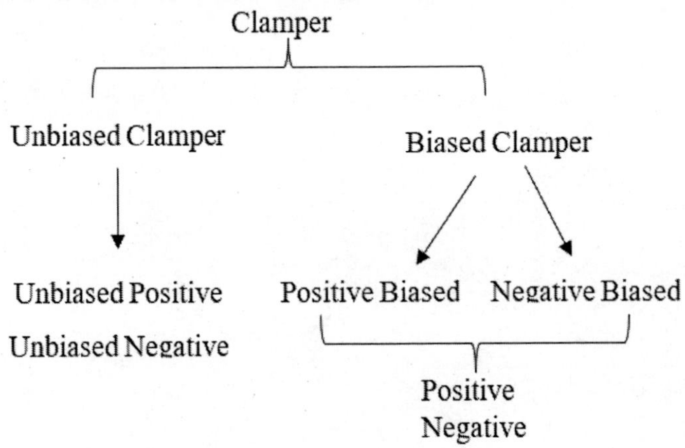


**Fig. 3.18:** Types of Clamper Circuit

## *Unbiased Positive Clamper*

The diode is connected to the load in parallel, as can be seen in Fig. 3.19. This means that the output at the load will be provided by the diode when it is in reverse bias.

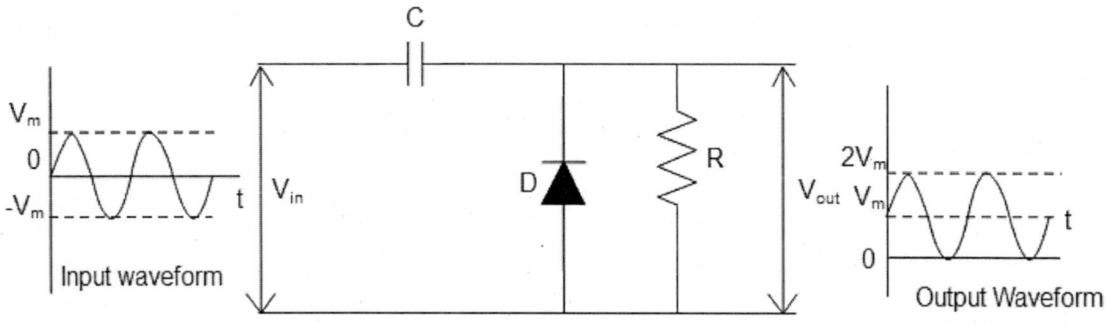


**Fig. 3.19:** Unbiased Positive Clamper

When the input signal is in its negative half cycle, the diode is forward biased, hence there is no output signal during that period. The capacitors charge to the peak value (Vm) of the input voltage Vin while the diode permits electric current to flow through it. The capacitor was charged in reverse polarity. The capacitor stores the charge till the diode remains forward biased when the input current falls after reaching its maximum value -Vin.

A signal arises at the output because the diode is reverse biased during the positive half of the input signal. When reverse biased, the diode does not allow electricity to pass through it. As a result, the input current flows directly to the output.

The diode is not conducting when the positive half cycle starts, and the capacitor's charge is released. As a result, the voltage at the output is equal to the sum of the input voltage ($V_{in}$) and the voltage stored in the capacitor ($V_m$).

$$V_0 = V_m + V_{in}$$

The output waveform above shows that the signal level is in this instance pushed upward or to the positive side. It is hence referred to as a positive clamper.

## *Unbiased Negative Clamper*

Negative clamper's operation in more depth by examining the Fig. 3.20.

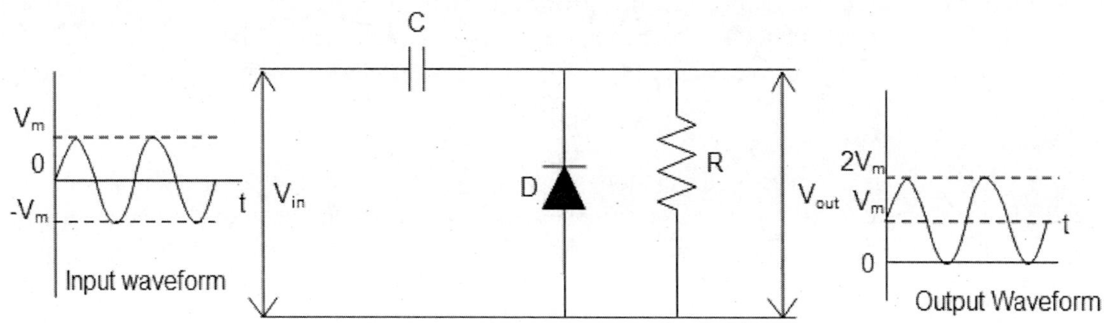


**Fig. 3.20:** Unbiased Negative Clamper

The diode is forward biased during the positive half cycle, hence there is no output signal. The diode permits current through it, when it is in a forward bias condition. These currents reverse-polarize the input signal and charge the capacitor to its peak input voltage ($-V_m$). The capacitor maintains the charge until the diode is forward biased as the current falls after reaching its maximum value ($V_{in}$).

When the negative half of the signal is supplied, the diode is now biased in the other direction. The circuit's output may now be utilized to find load current as a consequence. The capacitor discharged because the diode was not conducting. The input voltage ($V_{in}$) and the capacitor voltage ($-V_m$) are therefore mixed at the output. So, the result will be

$$V_0 = -V_m + V_{in}$$

## *Biased Positive Clamper*

In some of the applications, there is a need to clamp the signal for a particular voltage level only. In such cases, the biased clampers are useful. The biased positive clamper is classified into positive biased positive clamper and a negative biased positive clamper.

i. **Positive Biased Positive Clamper**

Although in this case an additional voltage is applied in order to cause a further change in the signal's level, the working is virtually identical to the unbiased positive instance.

When the input signal during the positive half cycle is lower than the battery voltage, the diode is forward biased by the battery voltage. The capacitor is charged by the battery voltage ($V_B$). The diode stops allowing current to flow through it when the input voltage exceeds the battery voltage because the diode becomes reverse biased.

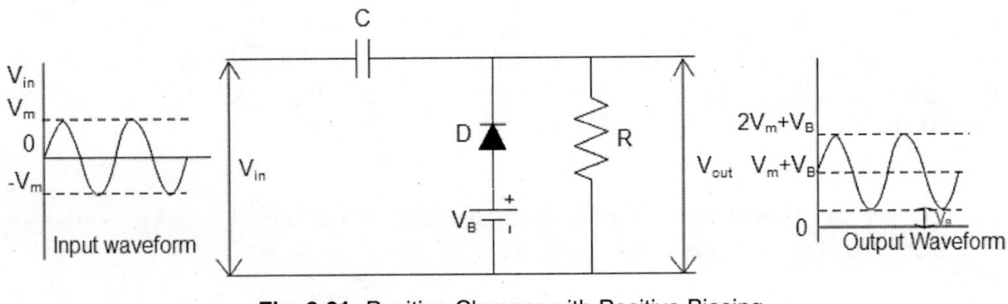


**Fig. 3.21:** Positive Clamper with Positive Biasing

The diode is now forward biased by both the input voltage and battery voltage when the polarity of the input signal is reversed. The diode, therefore, permits current. The capacitor will receive this current and be charged by an amount of input signal and battery voltage ($V_m + V_B$). From the next positive cycle, the input signal is added up with the capacitor charge and behaves like an unbiased positive clamper circuit.

ii. **Negative Biased Positive Clamper**

The battery voltage reverse biased the diode when the input voltage was higher than the battery voltage during the negative half cycle. As a result, the signal is produced at the output. When the input voltage is less than the battery voltage, the diode allows electric current to flow through it. Because the diode is forward biased by the input voltage. The capacitor will be charged by an amount of $V_m - V_B$.

The input voltage and battery voltage both reverse bias the diode during the positive half cycle. So, the output signal was equal to the sum of the input voltage and capacitor voltage.

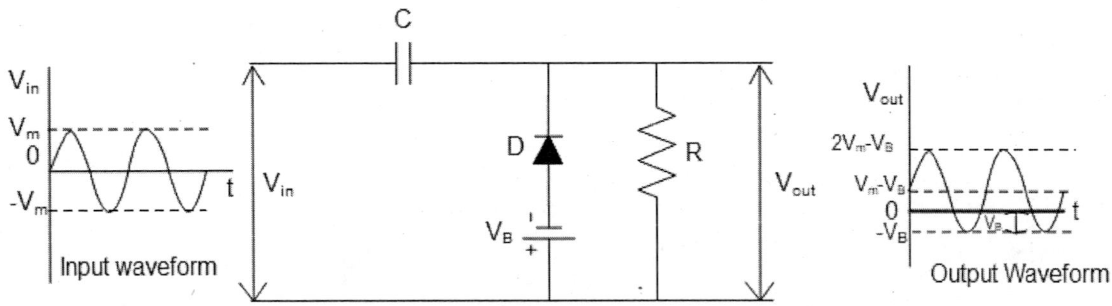


**Fig. 3.22:** Positive Clamper with Negative Biasing

## *Biased Negative Clamper*

Similar to before, the negative clamper circuit receives positive and negative biasing. Let's address each scenario independently as we move forward. The biased negative clamper is further classified into positive biased negative clamper and negative biased negative clamper.

i. **Positive Biased Negative Clamper**

The battery voltage (VB), reverse bias the diode when the input signal (Vin) is lower during the positive half cycle than the battery voltage. When the input supply voltage exceeds the battery voltage, the input signal voltage causes the diode, which is forward biased, to allow an electric current to flow through it. This current will reach the capacitor, which will then be charged by an amount of $-V_m + V_B$.

Input voltage and battery voltage both reverse bias the diode during the negative half cycle. So, the output is the sum of the input voltage and capacitor voltage.

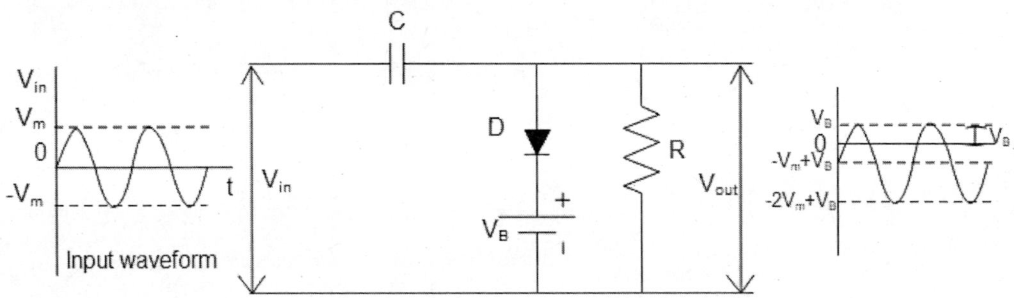


**Fig. 3.23:** Negative Clamper with Positive Biasing

ii. **Negative Biased Negative Clamper**

Both the input signal ($V_{in}$) and battery voltage ($V_B$) forward bias the diode during the positive half cycle. As a result, the capacitor is charged by an amount of $-V_m - V_B$, due to the reverse-polarized charge by the current flowing through it.

The battery voltage forward biases the diode when the input signal ($V_{in}$) exceeds the battery voltage during the negative half cycle. When the input signal ($V_{in}$) is less than the battery voltage, the diode is reverse biased by the input signal, which results in downward shift in the output signal.

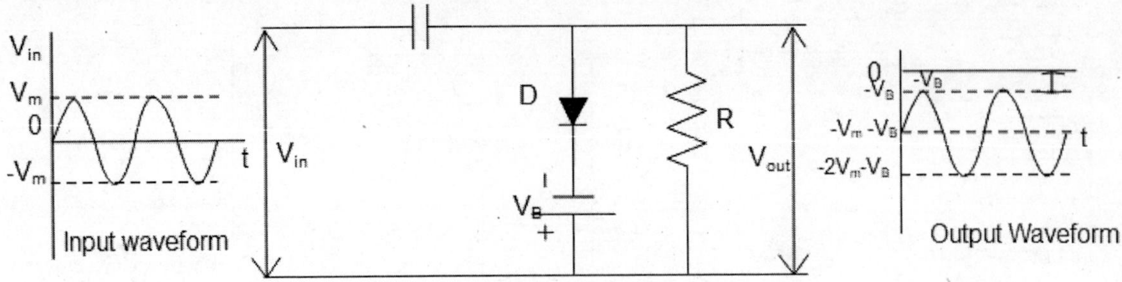


**Fig. 3.24:** Negative Clamper with Negative Biasing

**Example 3.3:** For the figure given, find 5τ. Compare 5τ to half to the period of the applied signal. Sketch $V_0$.

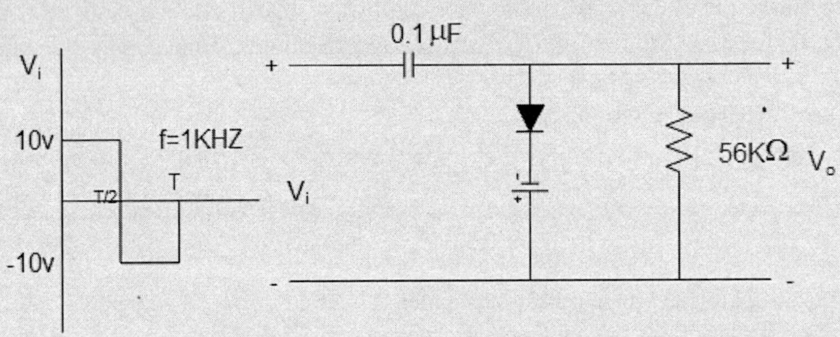


**Solution:**

$$\tau = RC = 56 \times 10_3 \times 0.1 \times 10^{-6}$$

$$= 5.6 \text{ ms}$$

$$5\tau = 5 \times 5.6 \times 10^{-3}$$

$$= 28\text{ms}$$

5τ to the half of the period of the applied signal.

$$T = \frac{1}{f}$$

$$T = \frac{1}{1 \times 10^3} = 1\text{ms}$$

Half time period will be, $\frac{T}{2} = \frac{1\text{ms}}{2} = 0.5\text{ms}$

The capacitor will fully charge or discharge in $5\tau$, $\tau$ must be large enough to ensure that the voltage across capacitor does discharge significantly during the interval when the diode off. It is inferred that, $5\tau \gg \frac{T}{2}$. So, discharging time of capacitor during the interval when the diode off.

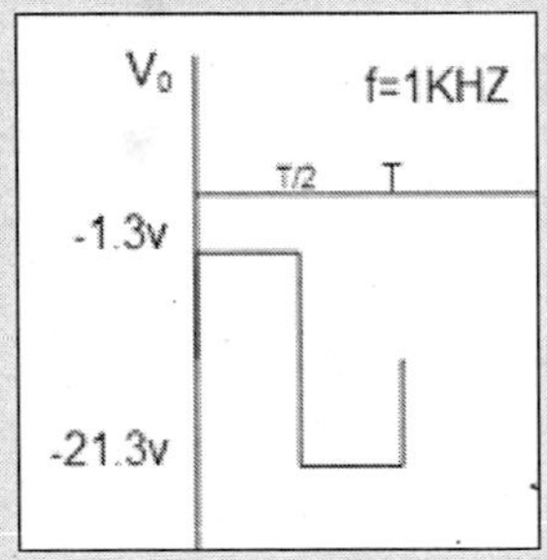


The output $V_0$ is,

During forward bias, the output $V_0 = 0.7 - 2 = -1.3V$

During the reverse bias, the output $V_c = 10 - 0.7 + 2 = 11.3V$

$$V_0 = -V_i - V_c = -10_V - 11.3V = -21.3V$$

## 3.3 ZENER DIODE AS REGULATOR

In the previous section, we saw how the P-N junction diode is operated in both the forward and reverse directions. We have also observed that there is hardly no current flowing through the diode during reverse bias conditions, where the applied voltage is lower than the breakdown voltage. However, as soon as the voltage across the diode exceeds this breakdown voltage, a significant amount of current begins to flow in the opposite direction. Therefore, this activity of the diode is referred to as the breakdown region. Additionally, this operating zone should be avoided for a typical diode. However, there are some diodes—known as Zener diodes—that are designed to be employed in this failure zone.

As a result, Zener voltage is the voltage at which the Zener diode functions in this breakdown zone and Zener current is the matching current that flows through the diode. Now, in contrast to a typical diode, the breakdown voltage of a Zener can range from 2V to 200V. However, as you can see, conventional diodes often had a breakdown voltage of greater than 20V. Therefore, in contrast to conventional diodes, these Zener diodes are extensively doped and the doping concentration is altered. The Zener voltage of the Zener diode is programmable. Therefore, the *Zener effect* is the main effect for Zener diodes whose breakdown voltage is less than 4V. The avalanche breakdown is the primary effect on the opposite end of the Zener diode whose breakdown voltage is more than 6V, and we have already covered these two processes in the previous section.

The below Fig. 3.25 shows the regulator application of the Zener diode. The circuit provides the constant DC voltage with respect to the change in input.

The Zener diode performs just like a standard all-purpose silicon PN junction diode when biased in the forward direction, or anode positive with regard to its cathode, passing the rated current.

The Zener diode, however, conducts in the opposite way once the reverse voltage reaches a certain level, unlike a typical diode.

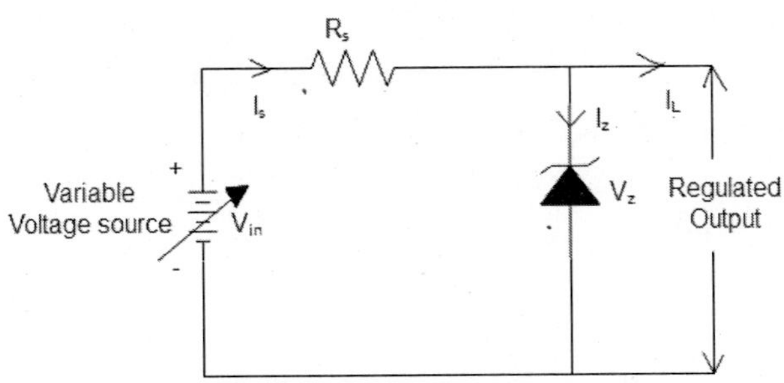


**Fig. 3.25:** Zener Regulator

When the cathode becomes more positive than the anode, it inhibits the flow of current through it.

Apply KCL, to analyse the circuit.

The Zener current is,

$$I_s = I_z + I_L$$

$$I_z = I_s - I_L$$

The amount of power passing through the Zener diode is,

$$P_z = V_Z I_Z$$

Scan to know more

**Example 3.4:** Find the Zener current and power dissipated across the Zener diode shown in the below figure.

**Solution:**

First, we need to ensure the operating region. In the given problem, to ensure the operating condition, calculate the Thevenin's equivalent voltage ($V_{th}$) across the diode.

The Thevenin's voltage is calculated as,

$$V_{th} = V_{in}\ (1k\Omega/1k\Omega+250\Omega)$$

$$V_{th} = 14.4V$$

The Thevenin's voltage ($V_{th}$) is greater than 10V. So, the diode is operated in breakdown voltage.

$$I_L = \frac{10V}{1K\Omega} = 10mA$$

$$I_s = \frac{18-10}{250} = 32mA$$

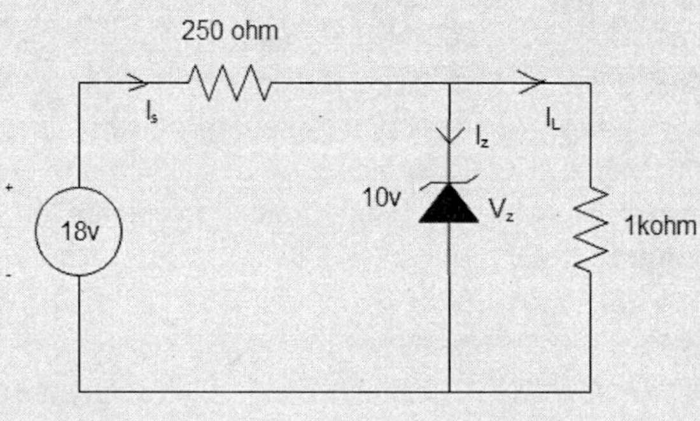


The diode current is,

$$I_z = I_s - I_L = 32 - 10 = 22mA$$

Power dissipated is,

$$P_z = I_z \cdot V_z = 22.10 = 220mW$$

## 3.4 APPLICATION OF SCHOTTKY DIODE

Schottky diodes are employed in fast-clamp diode applications as switches. Due to its improved frequency responsiveness, shorter switching times, and reduced power consumption, Schottky transistor-transistor logic (TTL), digital logic gates and circuits frequently use Schottky diodes.

### *Advantages*

Due to their low on-voltage, rapid recovery time, and minimal energy loss at higher frequencies, Schottky diodes are used. By enabling a fast transition from the conducting to blocking states, these features enable Schottky diodes to correct a current. So, in many situations, Schottky diodes are a great option for semiconductor devices.

Applications of Schottky diodes include,

- Radio Frequency mixer and detector
- Power rectifier
- Solar cell
- Clamper diode applications

Let's go through each of them individually below.

### *RF Mixer and Detector*

Schottky diodes are used in radio frequency applications due to their fast-switching speed and high frequency characteristics. The many metal-semiconductor junction designs of Schottky diodes also make these semiconducting components useful in power detectors or mixer circuits.

### *Power Rectifier*

Schottky diodes, as opposed to regular PN junction devices, offer a high current density and a low forward voltage drop, making them the ideal semiconductor devices for use in power rectifier applications. These benefits help to reduce heat levels, and reduce the size of the heat sinks employed in that design to increase the overall effectiveness of the electronic system.

### *Application of Power and Circuit*

Schottky diodes can be used when two parallel power supplies are used to generate the current. Schottky diodes offer characteristics that make them perfectly suited for use in power applications because of their minimal forward voltage drop. These diodes are there because they stop reverse current flow from one source going into another.

### *Solar Cell*

Because the sun is not a constant source of energy, rechargeable batteries are widely used in conjunction with solar cells to store energy. The diode of SiC Schottky stops high-performance solar cells from discharging via lower-performance solar cells and keeps batteries from discharging through solar cells at night.

### *Clamp Diode*

Fast clamp diode applications use Schottky diodes as switches. The base junction is forward biased in this application. Schottky diodes greatly reduce the turn-off time while accelerating the circuit speed.

## 3.5 APPLICATION OF LED

The main purpose of LEDs (Light Emitting Diodes) is to illuminate objects and even spaces. Because of its small size, low energy consumption, long lifespan, and ability to be employed in a wide range of applications, it is widely used.

### *Use of LEDs as TV Backlight*

The main source of energy consumption in a TV is the backlight. Utilizing LEDs helps to reduce power consumption effectively. Using an LED in the margins of the TV, will be a less expensive option. Better contrast may be achieved by placing LEDs immediately behind the monitor. When it comes to TV backlighting, LEDs have taken the place of Compact Fluorescent Lamps (CFLs) and Liquid Crystal Displays (LCDs).

### Smartphone Backlighting

The backlight design of the smartphone may be made slimmer and more affordable with the use of LED's. The cost of LED may change depending on how big the smartphone's screen is. They guarantee greater battery life because of the lower output voltage.

### Use of LED in Display

The use of LED display boards outside, such as for storage signs, billboards, and traffic signs has become common place in modern times. Signboards that display messages in many languages will use less energy if more LEDs are added.

### LEDs in Automotive

LED applications in the automobile sector are expanding. Energy is conserved and visibility is improved using LEDs. For greater accessibility, they are frequently utilized in the back and rear of a car. As LED lighting improves visibility while it is ON, OFF, or muted throughout any stage of the journey, it can increase both the safety of drivers and pedestrians.

### LEDs in Dimming Lights

A few LED applications allow for light dimming, which reduces energy usage. Appliances also employ this dimming function, which comes in two varieties. Global dimming, which dims all LEDs simultaneously.

### White PhlatLight LED

The SST-90 enables manufacturers of lighting fixtures to swap out LED arrays and bulbs for a single PhlatLight LED by combining the advantages of high power and efficiency. This leads to simpler designs, reduced costs, and a quicker time to market for LED clients.

Scan to know more

**Fig. 3.26:** White PhlatLight LED

### LED MR16 Replacement Lamps

One of the top manufacturers of LED MR16 halogen replacement lights, CRS Electronics, made a substantial performance advancement. 325 lumens of temperature-stabilized light output with a colour temperature of 2850K. More than four times as efficient as typical halogen MR16 lights, the light also features a colour rendering index (CRI) of 88, 1800 candlepower in the centre beam, and an efficiency of 53 lumens per watt.

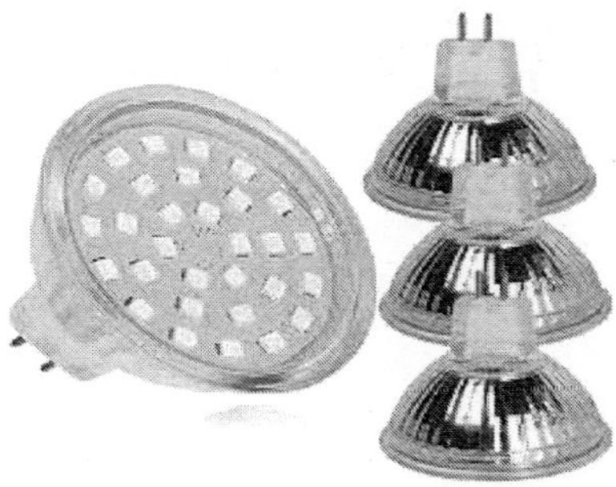

**Fig. 3.27:** LED MR16 Lamps

## 3.6 APPLICATION OF PHOTODIODE

Photodiodes are semiconductor devices that convert light energy into electrical current. They are widely used in various applications where light detection is required. Here are some common applications of photodiodes:

- Consumer electronics goods including CD's, alarms, equipment, and controls, all use photodiodes that are accustomed to operating, everything from air conditioners to televisions. Either photodiodes or photoconductors can be utilized for a variety of purposes. Both kinds of photosensors can be used to detect the presence of light, flash meters, or respond to it, for as by turning on the night time street lights.
- Electric isolation is provided by photodiodes with the help of optocouplers. When lit by light, optocouplers are used to optically connect two independent circuits. Optocouplers are faster than conventional devices in comparison.
- Photodiodes are used in safety equipment like smoke and fire detectors.
- Several medical applications for photodiodes. They are also utilized in blood gas monitors, computed tomography detectors, and machines that examine samples.
- Solar cell panels employ photodiodes.
- Logic circuits employ photodiodes.
- The detecting circuits employ photodiodes.
- Character recognition circuits employ photodiodes.
- In research and industry, photodiodes are used to precisely quantify light intensity.
- Since photodiodes are quicker and more complicated than a typical PN junction diode, they are commonly utilized in optical communication and lighting control.

### Alarm Circuit Using Photodiode

The employment of a photo-diode in an alarm system is depicted in Fig. 3.28. A photodiode installed in the entryway is exposed to light coming from a light source. As long as the light beam is unbroken, the reverse current ($I_R$) will keep flowing. The reverse current reduces to the dark current level when someone walks through the door, breaking the light beam. An alert is therefore triggered.

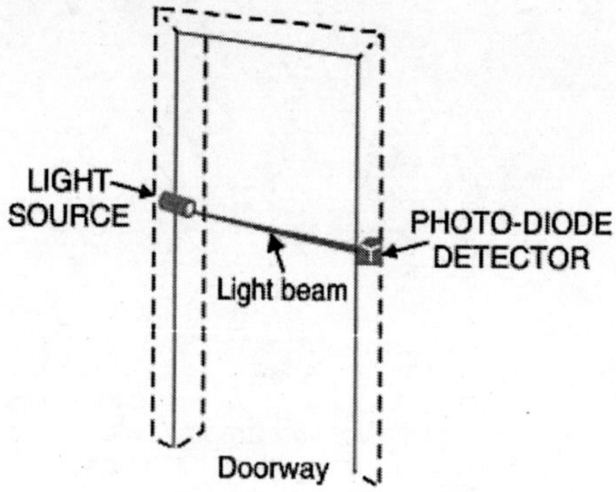


Scan to know more

**Fig. 3.28:** Alarm Circuit Using Photodiode

### *Photodiode in Counter Circuit*

A photodiode circuit utilized in a conveyor-based system for object counting is shown in Fig. 3.29. Using a photo-diode, you can count the items moving along a conveyor belt. In this circuit, a light source transmits a focused beam of light to a photodiode through a conveyor. The light beam is interrupted when the object passes, the IR level decreases to the dark current level, and the count goes up by one.

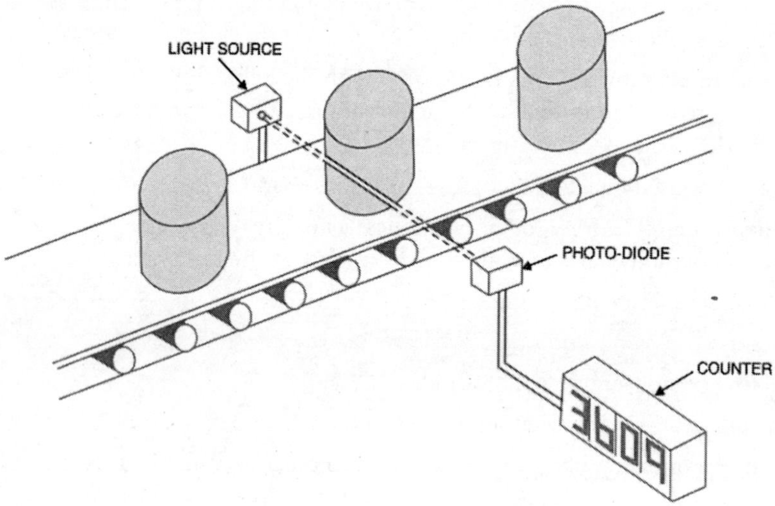


**Fig. 3.29:** Counter Circuit using Photodiode

## 3.7 APPLICATION OF SOLAR CELL

A solar cell, also known as a photovoltaic cell, is an electrical device that uses the photovoltaic effect, a natural physical and chemical phenomenon, to convert light energy directly into electricity. A device whose electrical properties, such as current, voltage, or resistance, change when exposed to light. The electrical building blocks of photovoltaic modules, sometimes referred to as solar panels, are frequently individual solar cell devices. The greatest open-circuit voltage that a typical silicon single junction solar cell can generate is between 0.5 and 0.6 volts.

### Exotic Places

It is not always possible, affordable, or even practical to extend power lines to locations where energy is needed. Rural residences, towns in underdeveloped countries, lighthouses, offshore oil platforms, desalination plants, and outlying medical facilities can all benefit from Photo Voltaic (PV).

### Stand-alone Power Applications

In urban or rural areas, PV can power stand-alone appliances, tools, and meters. PV can be used to power a variety of equipment, including parking meters, temporary traffic signs, emergency phones, radio transmitters, water irrigation pumps, stream-flow gauges, remote guard posts, highway lighting, and more.

### Power in Space

For many years, the primary energy source for spacecraft that orbit the earth has been solar power. Planetary and space exploration will continue to depend on high-efficiency photovoltaic technology. The International Space Station and surface rovers on the Moon and Mars have both benefited from its power.

### Building Related Needs

PV panels set up on the ground or on rooftops can provide power for structures. To accomplish a dual purpose, PV material can also be used to construct buildings as cladding, roof tiles, or windows. Additionally, parking garages and grass areas can be PV-covered to provide lighting and shade.

### Military Uses

Flexible, light-weight thin-film photovoltaics can be used in situations where mobility or toughness are essential. Lightweight PV is transportable by soldiers and may be used to power electronic devices in the field or at distant locations.

### Transportation

Vehicles like automobiles and boats may use solar energy as an additional power source. For onboard electricity or to slowly charge batteries, vehicle sunroofs may have PV. Lightweight PV may also adjust to an airplane wing's curvature in order to power high-altitude aircraft.

## UNIT SUMMARY

- Diode as a switch.
- Reverse bias–Resistance gets increased–Diode act as an open switch.
- Forward bias–Resistance gets reduced–Diode act as a closed switch.
- Clipper–Remove the portion of the signal.
- Clamper–Shift or clamp level of the signal.
- Schottky diode–short switching time, less power consumption, used for high frequency applications.
- Reverse bias operation for Zener diodes.
- The regulator is a Zener diode–Provides a constant voltage with respect to the input voltage.
- LED–Small size, low power consumption, long lifespan.
- Solar cell–It is also known as photovoltaic cell. It can be used for energy saving purpose.

# EXERCISES

## Multiple Choice Questions

3.1 The ideal diode is one which behaves as a

a. Conductor b. Resistor

c. Rectifier d. Clipper

3.2 What circuit is used to cut off the specified amount of the input signal?

a. Clipper b. Clamper

c. Diode d. LED

3.3 Which circuit is used for rectification purpose?

a. Clamper b. LED

c. Solar Cell d. Clipper

3.4 Name the circuit which is used to shift the signal that is applied on the input

a. Clamper b. Clipper

c. Diode d. LED

3.5 The purpose of positive clipper is

a. Clip off the negative portion of the signal b. Clip of the negative and positive cycle

c. Clip of the positive portion of the input d.

None of the above

3.6 To remove the negative portion of the input which type of clipper is used

a. Positive clipper b. Clamper

c. Positive clamper d. Negative clippere.

3.7 To shift the signal to the positive portion which type of circuit is used

a. Positive clamper b. .Negative clamper

c. Both clamper d. Clipper

3.8 The purpose of biasing circuit is to

a. Clip the particular portion of the input b. To remove the signal from the circuit

c. To add some portion to the input d. None of the above

3.9 Which is not a component of a clamper circuit

a. Capacitor b. Resistor

c. Diode d. Inductor

3.10 If the signal is need to shift 2V above the positive portion. What is its worth biasing network?

a. 2V b. 1V

c. 0V d. 5V

3.11 Radio frequency mixer circuit use

a. Schottky diode b. LED

c. Photodiode d. Solar cell

3.12 Which kind of circuit used for TV backlighting purpose

a. Schottky diode b. LED

c. Photodiode d. Solar cell

3.13 For regulator, the Zener diode should be operated at

a. Reverse bias  b. Ideal
c. Zero bias  d. Forward bias

3.14 Zener breakdown voltage will be in the range of

a. 20-200V  b. 10-20V
c. 0.1-10V  d. 1-10V

3.15 The purpose of combinational clipper circuit is

a. Clip small portion in the positive cycle
b. Clip the small portion in the negative cycle
c. Clip small portion in both positive and negative cycle
d. Clamp small portion

3.16 Types of clipper circuit is

a. 1  b. 4
c. 5  d. 8

3.17 Types of clamper circuit

a. 4  b. 5
c. 6  d. 1

3.18 When the diode is act as a closed switch

a. Forward bias  b. Reverse bias
c. Zero bias  d. None of the above

3.19 The diode is operated in reverse bias, then the diode acts like a

a. Closed switch  b. Open switch
c. SCR  d. LED

3.20 Diode is in forward bias, the resistance offered by the diode is

a. Infinite  b. Zero
c. Some finite value  d. None of the above

3.21 Reverse bias condition, resistance offered by the diode is

a. Infinite  b. Zero
c. Some finite value  d. None of the above

| Answer | | | | |
|---|---|---|---|---|
| 3.1 **c** | 3.2 **a** | 3.3 **d** | 3.4 **a** | 3.5 **c** |
| 3.6 **d** | 3.7 **a** | 3.8 **a** | 3.9 **d** | 3.10 **a** |
| 3.11 **a** | 3.12 **b** | 3.13 **a** | 3.14 **a** | 3.15 **c** |
| 3.16 **d** | 3.17 **a** | 3.18 **a** | 3.19 **b** | 3.20 **b** |
| 3.21 **a** | | | | |

## SHORT AND LONG ANSWER TYPE QUESTIONS

### Category I

3.1 Explain the working of diode in forward bias condition.

3.2 Discuss the switching mechanism of diode.

3.3 What is regulator.

3.4 Explain the breakdown mechanism of Zener diode,

3.5 What is clipper and its types.

3.6 What is clamper and its types.

3.7 Explain biasing,

3.8 Sketch the diagram for positive clipping.

3.9 Draw and explain the clamper with biasing circuit.

3.10 What is photodiode?

3.11 List out the application of solar cell.

3.12 List out the application of LED.

3.13 Give the applications of Schottky diode.

### Category II

3.1 Discuss the operating principle of diode in both forward and reverse bias condition with suitable diagram.

3.2 Explain in detail about clipper and its types using the block diagram.

3.3 Discuss about the clamper circuit. Illustrate the types of clampers with neat diagram.

3.4 Enumerate the working principle of the regulator using Zener diode.

3.5 Explain in detail about the application of solar cell, photodiode, and LED.

### Numerical Problems

3.1 Sketch the output for the diagram given below.
**(Ans: Forward bias $V_0 = 0$, Reverse bias $V_0 = 2V$)**

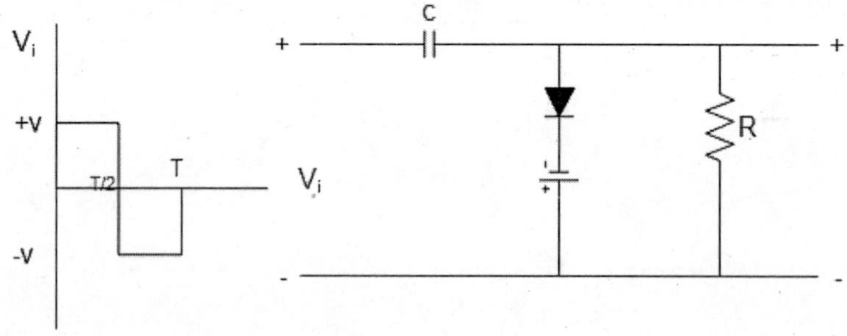


3.2 Calculate the time constant for problem 3.1.
**(Ans: $\tau = 0$)**

3.3 Consider the example 3.3. Find 3τ. Compare 3τ to the applied signal has the frequency of 2KHz.

**(Ans: 3τ = 16.8ms, T = 2ms)**

3.4 Calculate the charging and discharging time of the circuit shown below.

**(Ans: $\tau_c = 0$, $\tau_{dc} = 50ms$)**

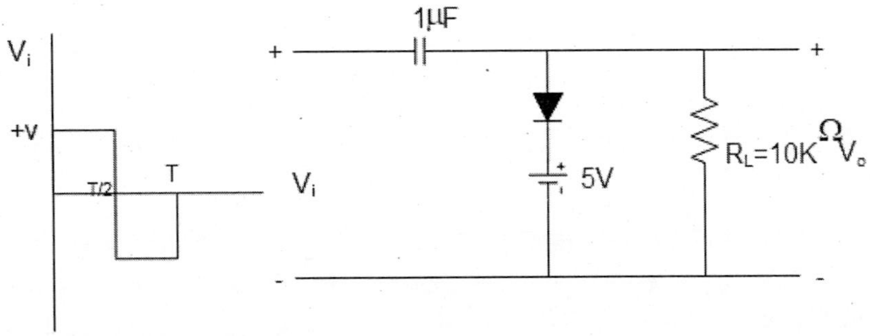


3.5 Determine the $V_0$. Consider Si diode has the $V_i = 20V_p$ and bias voltage of 10V.

**(Ans: forward bias $V_0 = 0.7V$, Reverse bias $V_0 = -20V$)**

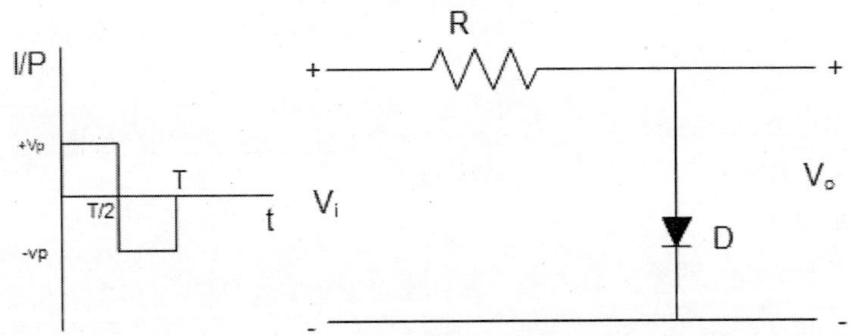


## PRACTICAL

## Photodiode and LED Characteristics

### *Aim*

i. To research the qualities of a light-emitting diode.

ii. To study the photodiode I-V properties.

### *Apparatus Required*

- LED, photo diode
- Ammeter (0-50mA)
- Voltmeter (0-10 V)
- DC power supply
- Rheostat

## Circuit Diagram

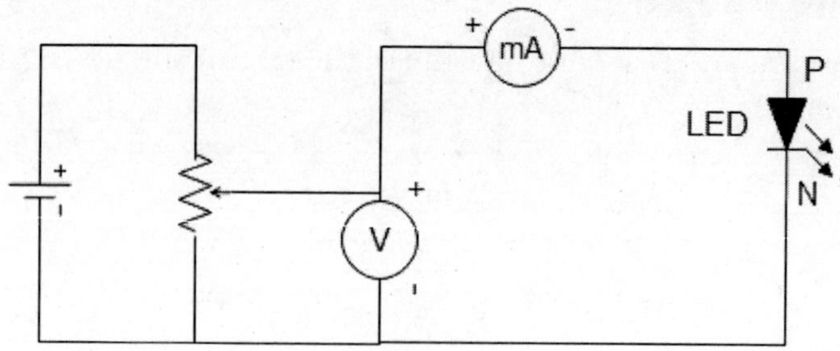


## Procedure

- According to the circuit schematic, connect the parts.
- Activate the power supply. A milliampere is used to gauge the current passing through the LED while the voltage is set to 0 V.
- The voltage is raised using the rheostat slider in increments of 0.2 V.
- The microammeter records the associated current for each adjustment of the voltage. In a table, the observations need to be listed.

## Observation

| Forward Voltage (Volts) | Forward Current (mA) |
|---|---|
| 0 | |
| 0.2 | |
| 0.4 | |
| 0.6 | |
| 0.8 | |
| 1.0 | |
| 1.2 | |
| 1.4 | |
| 1.6 | |
| 1.8 | |

## Photodiode Characteristics

It is a PN junction diode made of silicon or germanium that is photosensitive under conditions of reverse bias, i.e. The amount of light exposure or light intensity affects the reverse current in a photodiode. This diode's applicability is when it is reverse biased. The voltage-current relation,

$$I = I_0\left(e^{\frac{V}{\eta v T}} - 1\right) - I_P$$

The reverse biased photo diode current equation will be,

$$I_R = -I_0 - I_P$$

## Circuit Diagram

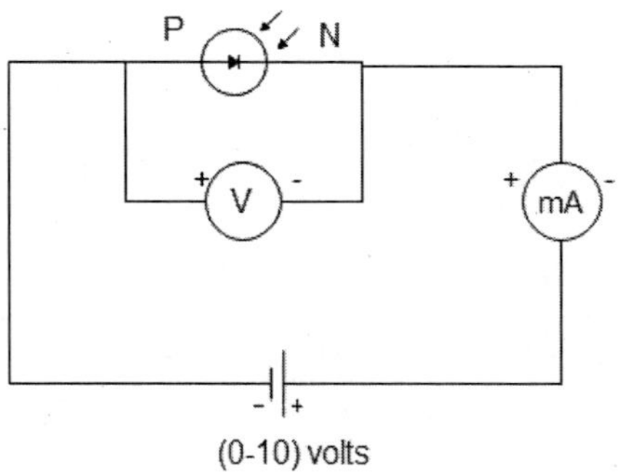


## Observation

### *Intensity of Light I=0*

| Reverse Voltage (Volts) | Reverse Current (mA) |
|---|---|
| 0 | |
| 0.2 | |
| 0.4 | |
| 0.6 | |
| 0.8 | |
| 1.0 | |
| 1.2 | |
| 1.4 | |
| 1.6 | |
| 1.8 | |

### *Result*

i. The characteristics of an LED are comparable to those of a forward PN-junction. For LEDs, the cut-in voltage, or the voltage at which conduction starts, is volts.

ii. The current-voltage characteristics of photodiode is analysed

## KNOW MORE

A semiconductor diode known as an LED, or light-emitting diode, illuminates in proper bias situation. Your gadgets, modern lights, and digital television monitors all require these components.

## History

Henry Joseph Round, a British radio researcher and Guglielmo Marconi's assistant, made the discovery of electroluminescence—a natural phenomenon upon which LED technology is based—in 1907 while working with silicon carbide and a cat's whisker.

Russian radio scientist Oleg Vladimirovich Lossev was researching electroluminescence in radio set diodes during the 1920s. He detailed his research in an article titled "Luminous Carborundum [silicon carbide] Detector and Detection with Crystals" that was published in 1927. Although his work did not directly lead to the development of the LED at that time, it did have an impact on subsequent innovators.

A decade later, in 1961, Robert Biard and Gary Pittman developed and patented an infrared LED for Texas Instruments. Since the original LED was an infrared gadget, it was invisible to the human eye. Humans cannot perceive infrared light. Ironically, Baird and Pittman only accidentally produced a light-emitting diode while trying to make a laser diode.

**Robert Biard**

## Organic LED

A light-emitting diode (LED) with an organic compound film acting as the emissive electroluminescent layer and producing light in response to an electric current is known as an organic light-emitting diode (OLED or organic LED) or organic electroluminescent (organic EL) diode. This organic layer is sandwiched between two electrodes, with at least one transparent electrode.

OLED displays don't need backlights because they instead emit visible light. It may therefore exhibit profound black depths and be smaller and lighter than a liquid crystal display (LCD). An OLED panel may be able to attain a greater contrast ratio than an LCD in low ambient light situations (such as a dark room), regardless of whether the LCD employs cold cathode fluorescent bulbs or an LED backlight. OLED displays are made in a manner similar to LCDs, but after the formation of the TFT (for active matrix displays), addressable grid (for passive matrix displays), or ITO segment (for segment displays), the display is coated with hole injection, transport, and blocking layers as well as with electroluminescent material. The entire stack of materials is then enclosed, with ITO or metal acting as the cathode once more.

Scan to know more

The TFT layer, addressable grid, or ITO segments are connected to or used as the anode, which may be made of ITO or metal. OLEDs may be made transparent and flexible, both display types are used in smartphones with optical fingerprint scanners and foldable devices.

## REFERENCES AND SUGGESTED READINGS


1. David.J.Griffiths, *Introduction to Quantum Mechanics,* Pearson Prentice Hall, 2005.
2. Chenming Calvin Hu, *Modern Semiconductor Devices for Integrated Circuits,* Pearson Prentice Hall, 2009.
3. S.O.KASAP, *Principle of Electronic Materials & Devices,* MC Graw Hill Education, 2018.
4. S.Datta, *Quantum Transport:Atom to Transistor,* Cambridge University Press, 2005.
5. Robert F. Pierret, *Semiconductor Device Fundamentals,* ADDISON-WESLEY Publishing Company, 1996.
6. Donald A. Neamen, *Semiconductor Physics and Devices Basic Principle,* MC Graw Hill Companies, 2012.
7. Ben G, Streetman, Sanjay Kumar Banerjee, *Solid State Electronic Devices,* Pearson Education Limited, 2016.
8. *https://physicsinmyview.com/2020/09/facts-about-quantum-physics.html*
9. *https://onlinecourses.nptel.ac.in/noc21_ee80/course*
10. *https://onlinecourses.nptel.ac.in/noc21_ee59/course*
11. *https://onlinecourses.nptel.ac.in/noc21_ee19/course*

# 04 BIPOLAR JUNCTION TRANSISTOR

## UNIT SPECIFICS

*We have dealt with the following topics throughout this unit:*

- *Evaluation of the junction transistor*
- *Construction of the Bipolar transistor*
- *Working principle of the junction transistor*
- *Analyse the various type of configuration of the BJT*
- *Study the input and output characteristics of the BJT*
- *Problems including operating point*
- *Concept of DC load line*

*The concepts of real-world applications are covered in order to increase creativity and curiosity as well as ability to solve problems.*

*The unit includes assignments through a number of numerical problems, a list of references, and suggested readings so that one can go through them for practice in addition to providing a big number of multiple-choice questions and questions with short and lengthy answer types designated in two categories following lower and higher order of Bloom's taxonomy. Inside the unit some QR codes have been given for getting added details on various topics of interest to be scanned for relevant supportive knowledge.*

*Following the content-based relevant practical, there is a "Know More" section. This subsection has indeed been thoughtfully created so that the additional material supplied will be useful to the book's readers. This module primarily spotlights the initial activity, examples of some interesting titbits, analogies, background of the advancement of the subject focusing on the key observations and findings, timelines beginning with the development of the concerned topics up to the present, case studies involving environmental, sustainability, social, and ethical issues, if appropriate, and lastly the unit's curiosity and inquisitiveness subjects. Applications of the subject matter for our everyday real life or/and industrial applications on a range of factors.*

## RATIONALE

*This unit on Bipolar Junction Transistor helps students to get a primary idea about the construction and physics behind the transistor. It explains operation and its various configurations. All these basic aspects are relevant to study the transistor properly. It then clearly explains the configuration of transistor and its input and output characteristics as well as relation between three different configurations. All these are discussed at length to develop the subject. Some related problems are pointed out for clear understanding, which can help further for getting a clear idea of the concern topics on Junction Transistors.*

*Transistor is an important branch of semiconductor device that essentially deals with various application in commercial world. BJT started its journey by study of diode forward and reverse bias characteristics of diode. This permits one to analyse the operation of many day-to-day familiar phenomena around us. But at the same time, it covers the stability of the transistor and importance of Q point. Its practical applications are related to the construction, design, and operation of different devices.*

## PRE-REQUISITES

*Basic concepts of PN Junction Diode*

## UNIT OUTCOMES

*List of outcomes of this unit is as follows:*

*U4-O1: Describe the construction of BJT*

*U4-O2: Working of the Bipolar Junction Transistor*

*U4-O3: Explain the various configuration of BJT*

*U4-O4: Describe the relationship among the base, emitter and collector currents of the BJT*

*U4-O5: Analyse the DC load line for stability*

*U4-O6: Realize the role of operating point*

| Unit-4 Outcomes | EXPECTED MAPPING WITH COURSE OUTCOMES (1- Weak Correlation; 2- Medium Correlation; 3- Strong Correlation) | | | | | |
|---|---|---|---|---|---|---|
| | CO-1 | CO-2 | CO-3 | CO-4 | CO-5 | CO-6 |
| U4-O1 | 3 | 3 | - | 3 | 3 | 1 |
| U4-O2 | 1 | 1 | 2 | 2 | 1 | - |
| U4-O3 | 2 | 1 | - | 3 | 2 | 1 |
| U4-O4 | - | - | 2 | 3 | 2 | 2 |
| U4-O5 | 3 | 3 | - | 3 | 3 | 1 |

## 4.1 BIPOLAR JUNCTION TRANSISTOR

In the previous chapter, we have discussed the semiconductor diodes and their applications. In this chapter, we shall focus our attention on the Bipolar Junction Transistor (BJT). A bipolar junction transistor is a three-terminal semiconductor device in which the operation depends on the interaction of majority and minority carriers hence it is named as a bipolar device. It consists of two p-n junctions which can amplify or magnify an input signal. The forward biased junction has a low resistance path whereas a reverse biased junction has a high resistance path i.e., signals are transferred from low resistance circuit to high resistance circuit. Therefore, the transistor transfers a signal from low resistance to high resistance. The prefix '*trans*' means the signal transfer property of the device while '*istor*' classifies it as a solid element in the same general family with resistors.

### 4.1.1 Terminals

A bipolar junction transistor consists of three regions:

- Emitter
- Collector
- Base

**Emitter**-Emitter is more heavily doped than any of the other regions because its main junction is to supply the majority carriers to the base. The emitter terminal is always forward biased with respect to the base so that it can supply a large number of majority carriers.

**Collector**-It forms the right-hand side section on the transistor and its main function is to collect the majority charge carriers coming from the emitter and passing through the base. In most transistors, collector region is made physically larger than the emitter region. Because during the transistor operation, much heat is produced at the collector junction and it has to dissipate much greater power.

**Base**-It forms the middle section of the transistor. It is very thin as compared to either the emitter or collector and is very lightly doped.

### 4.1.2 Types

A transistor has three doped semiconductor regions and has two distinct structural types.

a. n-p-n transistors
b. p-n-p transistors

The p-type semiconductor material is placed between two n-type semiconductors to form a npn transistor as shown in Fig. 4.1 (a). An n-type semiconductor is sandwiched between two p-type semiconductors in a pnp transistor shown in Fig. 4.1 (b). The two junctions are emitter-base junction (J1) and base-collector junction (J2).

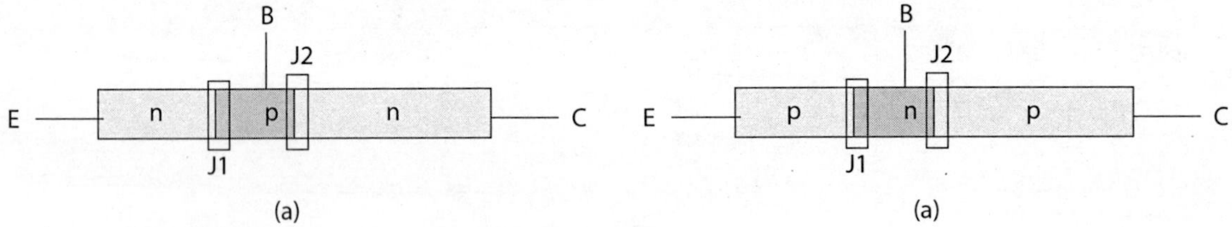


**Fig. 4.1:** Types of Bipolar Junction Transistor: (a) npn, (b) pnp

## 4.2 REGION OF OPERATION

The different operating regions of the Bipolar Junction Transistor is shown in the Table 4.1. In the active region, the transistor operates as an amplifier. The base-emitter junction is forward-biased, while the base-collector junction is reverse-biased. In the cut-off region, both the base-emitter and base-collector junctions are reverse-biased. In this condition, the transistor is effectively turned off, and no current flows between the collector and emitter. In the saturation region, both the base-emitter and base-collector junctions are forward-biased. In this condition, the transistor operates as a switch and allows a significant current to flow between the collector and emitter. The reverse active region is a less common operating condition for BJTs. In this region, the base-emitter junction is reverse-biased, while the base-collector junction is forward-biased.

**Table 4.1:** Different Operating Conditions of BJT

| *Junction 1* | *Junction 2* | *Region of Operation* | |
|---|---|---|---|
| Forward Bias | Reverse Bias | Active region–Amplifier | |
| Forward Bias | Forward Bias | Saturation–"ON" | Switch |
| Reverse Bias | Reverse Bias | Cut-off–"OFF" | |
| Reverse Bias | Forward Bias | Reverse – active region | |

These operating conditions are important to understand and control in BJT circuits, as they determine the transistor's behavior and its ability to amplify signals or switch currents. The biasing arrangement and applied voltage levels are crucial in determining the operating region of a BJT.

## 4.3 SYMBOL

The transistor symbol is represented by a circle that holds a vertical line representing the base terminal. The emitter terminal has an arrow symbol to the line that differentiates npn and pnp transistors with respect to the current flow. The remaining line without an arrow indicates the collector terminal.

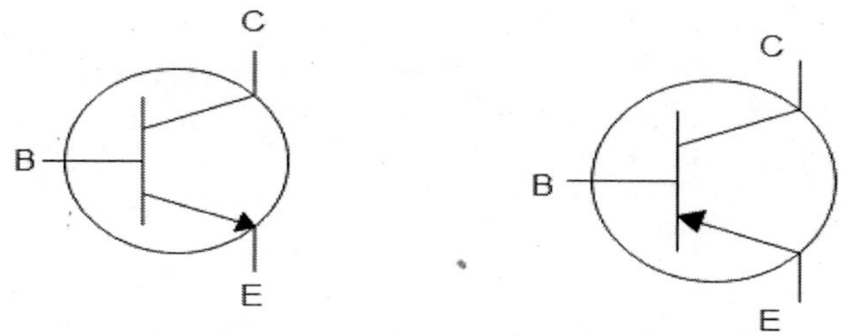


a) N-P-N b) P-N-P

(a) (b)

**Fig. 4.2:** Symbol of BJT: (a) npn, (b) pnp

For NPN connection, current flows from the base to the emitter as indicated by an outgoing arrow shown in Fig. 4.2 (a). whereas, for a PNP connection the current passes from the emitter to the base as indicated by an inward arrow shown in Fig. 4.2 (b). NPN transistors are more popular because the electron mobility is high compared to the hole mobility.

## 4.4 OPERATION OF NPN TRANSISTOR

The BJT is biased by applying appropriate voltages to the three terminals: the emitter, the base, and the collector. Biasing establishes the operating conditions and regions of the transistor. Fig 4.3 shows the npn transistor with forward bias applied at emitter-base junction and reverse bias is applied at collector-base junction. Due to the forward bias, the potential barrier at the junction is reduced and the electrons from the n-type emitter to flow towards the base. This constitute the emitter current $I_E$ As these electrons flow through p-type base, few electrons combine with the holes in the base region. As the base is thin and lightly doped small amount of recombination will take place in the base. Most of the electrons from the emitter passes over the collector region. For example, N number of electrons enter the base, out of which $(1-\alpha)$ $N$ electrons combine with the holes in the base region constitute a base current $I_B$ and $\alpha N$

electrons move to the collector region constitute a collector Current $I_C$. Only 2 to 5% of electrons are combined in the base and 95% to 98% of electrons are collected in the collector region. The emitter current is the sum of base current and collector current i.e,

$$I_E = I_B + I_C \tag{4.1}$$

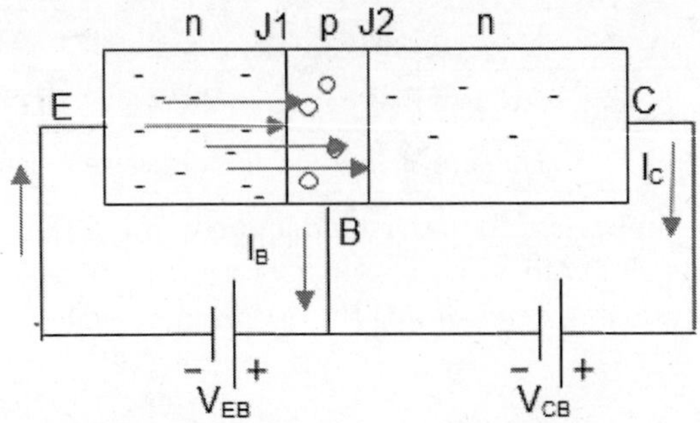


**Fig. 4.3:** Operation of npn Transistor

When the transistor is operated in no biasing condition, the barrier potential is same at the Junction J1 and J2. When the biasing is added to the transistor, the potential barrier varies at both junctions. Let's examine the motion of electrons and holes. The applied bias voltage $V_{EB}$ decreases the potential at junction J1. Most of the electrons from emitter cross over junction J1 and move to the base and recombine with the holes in the base region.

In practice, reverse saturation current starts flowing from the collector to the emitter when junction J2 is reverse biased. This current is also called leakage current denoted by $I_{CO}$. The current through the collector is given by, $I_c = \alpha I_E + I_{CO}$

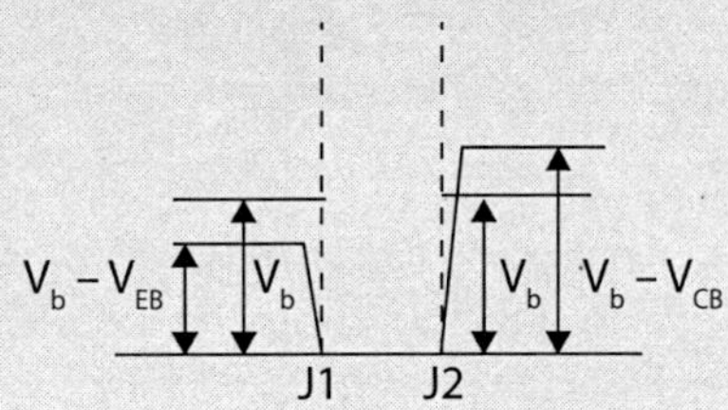


## 4.5 OPERATION OF PNP TRANSISTOR

In a pnp transistor the emitter-base junction is forward biased and collector-base junction is reverse biased. The holes in the p-type emitter to flow towards the base. This causes the emitter current IE As the holes crosses into n-type base region, the width of the base region is very thin and lightly doped, the small amount of holes recombine (less than 5%) with the electrons in the n-type base. This constitutes a base current $I_B$. The remaining holes are able to drift across the base and enter the collector region to constitute collector current $I_C$

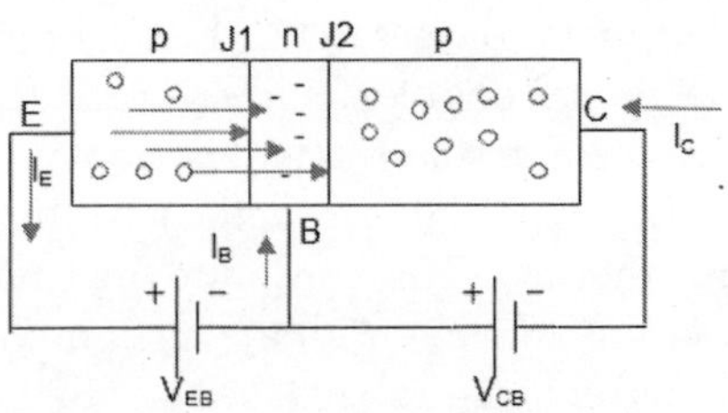


**Fig. 4.4:** Operation of pnp Transistor

It is noted that current conduction in npn transistor is by holes. Besides of hole current, there is electron current which flows from the base to the emitter region. This current depends upon emitter base potential. As the width of base region is very small, the ratio of hole current to the electron current is very small. So, for all practical purpose, electron current can be neglected. Thus, only hole plays a very important role in the operation of pnp transistor.

## 4.6 TYPES OF BJT CONFIGURATION

A transistor can be connected in a circuit in the following three possible ways.

i. Common Base configuration

ii. Common Emitter configuration

iii. Common Collector configuration

Each circuit connections have specific advantages and disadvantages. It may be noted here that regardless of circuit connection, the emitter is always biased in the forward direction, while the collector always has a reverse bias. Here, npn transistor is used to analyse these configurations.

### 4.6.1 Common Base Configuration

In the common base configuration, the base terminal of the transistor is common to the input and the output. A weak signal is introduced at the input and the amplified signal is obtained at the output of the transistor. Let us consider the device operated in an active mode of operation.

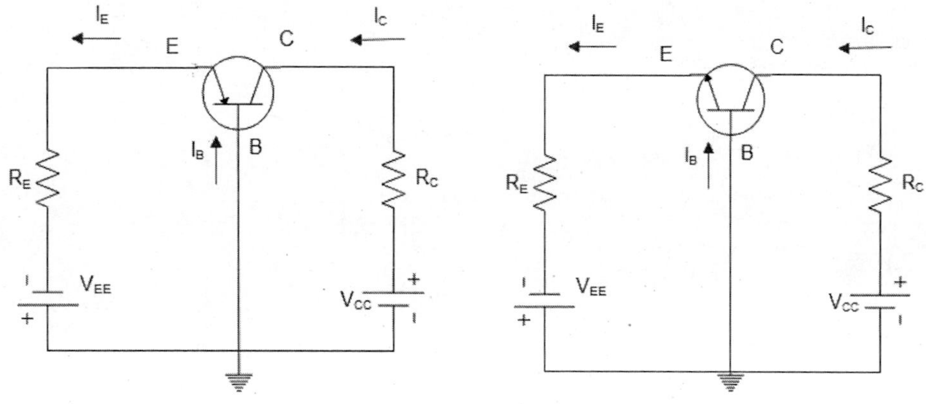


(a) (b)

**Fig. 4.5:** Common Base Configuration: (a) npn Transistor, (b) pnp Transistor

The resistance $R_E$ is the input resistance and $V_{EE}$ is the forward biasing potential. On the other hand, the reverse bias potential is $V_{CC}$ and the resistance is $R_C$. In case of transistor input and output characteristics are separately analysed, the input characteristics are plotted between the input current ($I_C$) and the input voltage ($V_{BE}$). Output characteristics are plotted between output current ($I_C$) and the output voltage ($V_{CB}$). The circuit is analysed by the Kirchoff's Current Law,

The emitter current is given by,

$$I_E = I_B + I_C \tag{4.2}$$

The collector current is given by,

$$I_c = \alpha I_E + I_{CBO} \tag{4.3}$$

$I_{CBO}$ is the reverse saturation current measured when the emitter terminal is open circuited. Since the minority charge carriers are relatively few in number. This results in a reverse saturation current. The emitter current is greater than the reverse saturation current. So, we can neglect the reverse saturation current ($I_{CBO}$). Now, the collector current will be,

$$I_C = \alpha I_E \tag{4.4}$$

The term α is rewritten as,

$$\alpha = \frac{I_c}{I_E} \tag{4.5}$$

The term α is called common base current gain. It is defined as the ratio of the output current to input current, and it is also called an amplification factor. The value α ranges between 0.95 to 0.98. This implies that 95% to 98% emitter current will be collected in the collector terminal. There is 5% to 2% recombination occurring in the base terminal.

**Example 4.1:** Determine the value of base current of a common base configuration whose amplification factor is 0.9 and emitter current is 1mA.

**Given:** $\alpha = 0.9$, $I_E = 1\text{mA}$

**Solution:** The base current $I_B = I_E - I_C$

We know that, $I_C = \alpha I_E$

$$I_c = 0.9 \times 1 \times 10_{-3}\ 0.92\ \text{mA}$$

$$I_B = (1\times10_{-3}) - (0.9 \times 10_{-3}) = 0.08\text{mA}$$

**Example 4.2:** In a common base connection, the emitter current is 1mA, $I_{CBO} = 50\mu A$, $\alpha = 0.92$. Find the total collector current.

**Given:** $\alpha = 0.92$, $I_E = 1\text{mA}$, $I_{CBO} = 50\mu A$

**Solution:**

Total collector current $I_C = \alpha I_E + I_{CBO}$

$$= (0.92 \times 1 \times 10_{-3}) + (50 \times 10_{-6})$$

$$= 0.97\text{mA}$$

## Input Characteristics

Consider the device operated in the active region of operation. Junction J1 is forward biased and junction J2 is reverse biased. On the input side, the current is $I_E$ and the voltage is $V_{EB}$. On the output side, the current is $I_C$ and the voltage is $V_{CB}$. The graphical representation, drawn between the input current ($I_E$) and input voltage ($V_{BE}$) at constant collector-base voltage. Fig. 4.6 shows the input characteristics of the common base configuration.

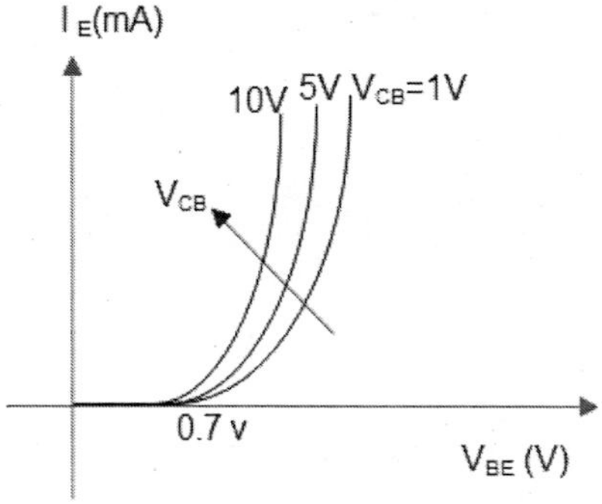


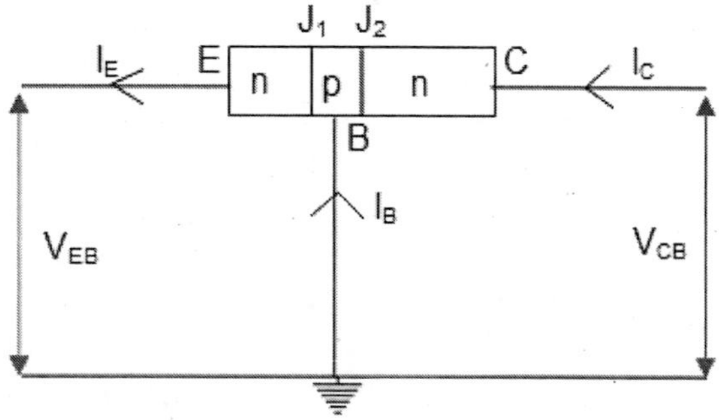


**Fig. 4.6:** Input Characteristics of CB Configuration

To determine the input characteristics initially, the output voltage $V_{CB}$ is set as zero, then $V_{EB}$ is increased. When $V_{CB}$ is zero, the emitter current $I_E$ increases rapidly with small change $V_{EB}$, it means that input resistance is very small.

The input resistance is defined as the ratio of change in emitter-base voltage to the change in emitter current.

$$\text{Input resistance}, r_i = \frac{\Delta V_{BE}}{\Delta I_E} \bigg| V_{CB = \text{constant}} \tag{4.6}$$

## Output Characteristics

It is the curve between collector current IC and the collector-base voltage VCB at constant emitter current $I_E$ Fig. 4.7 shows the output characteristics of a common base configuration. The output voltage $V_{CB}$ is increased from zero, the collector current $I_c$ varies at very low voltages (<1V). when the value of $V_{CB}$ is raised above 1-2 V, the collector current becomes constant. It means that, $I_C$ is independent of $V_{CB}$ and depends only on $I_E$.

The output resistance is defined as the ratio of change in collector-base voltage to the change in collector current.

$$\text{Output resistance}, r_i = \frac{\Delta V_{CB}}{\Delta I_C} \bigg| I_{E = \text{constant}} \tag{4.7}$$

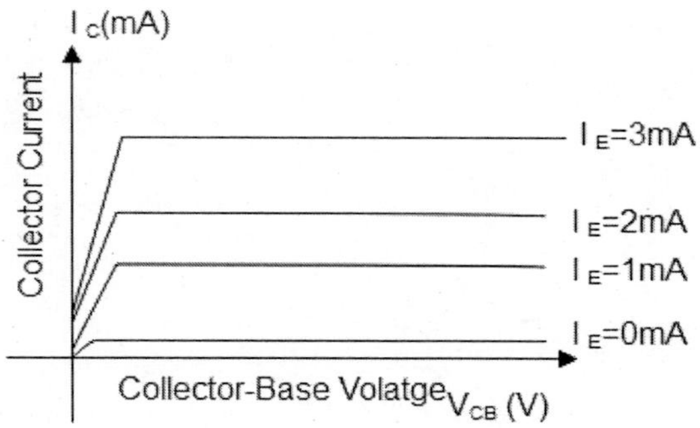


**Fig. 4.7:** Output Characteristics of Common Base Configuration

## Early Effect

It is named after the author James M early who discovered the effect. This effect is also known as *Base Width Modulation.* There is a modulation in the base width when the reverse bias voltage $V_{CB}$ is increased. This is because the base area is only faintly doped, the depletion layer widens under conditions of reverse bias at Junction J2 and penetrates deeper. The concentration of the hole is very low, so easy uncovering of immobile ions takes place. Depletion layer width is denoted as $W_B$ and it is called width of the base. Base width is also known as *metallurgical base width.* Let W be the width at which the depletion region reaches the base region.

The base width is provided as,

$$W_B = W_{eff} + W$$

The effective width is obtained as,

$$W_{eff} = W_B - W$$

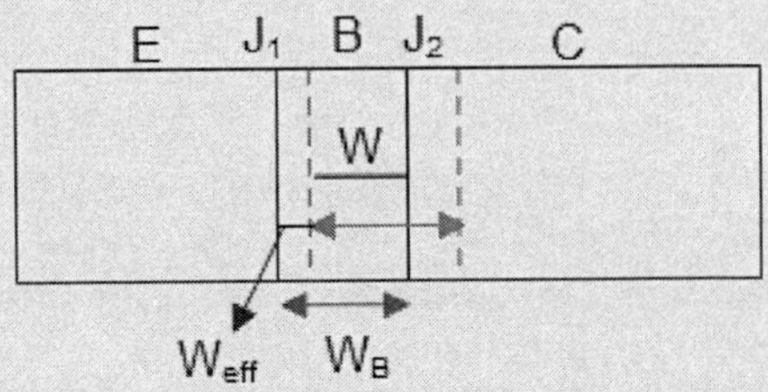


Increasing the reverse bias potential $V_{CB}$ will increase the width (W). The increased value of W will decrease the effective width $W_{eff}$ where the recombination takes place, and the width of this region is equal to the effective width. The base width reduction leads to the reduction of the recombination of carriers, hence the current $I_E$ gets increased. This generates a concentration gradient between the emitter and the base. Now, more electrons will move into the collector region.

### 4.6.2 Common Emitter Configuration

The input is applied between the base and the emitter, and the output is obtained between the collector and the emitter in a common emitter configuration. In this, the input and output circuits share the transistor's emitter, giving rise to the term "common emitter configuration". Fig. 4.8 depicts the configuration for both the npn and the pnp transistors.

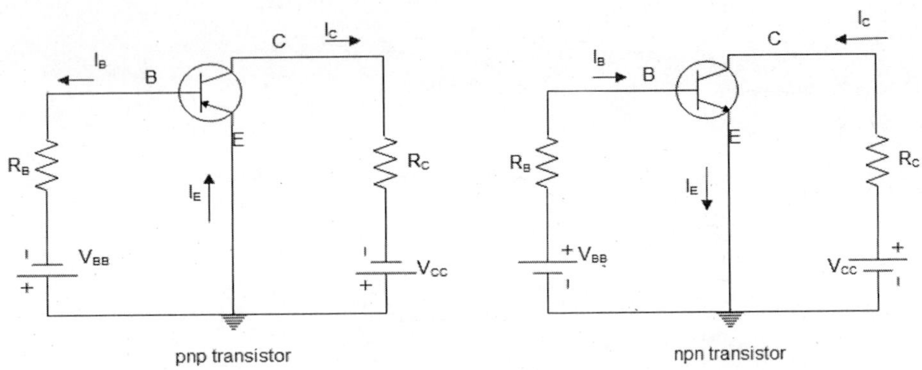


**Fig. 4.8:** Common Emitter Configuration

In common emitter configuration, the graph is drawn between the base-emitter voltage ($V_{BE}$) and base current ($I_B$) for varied output voltage ($V_{CE}$).

In CB configuration we have seen that,

$$I_c = \alpha I_E + I_{CBO}$$

$$I_C - I_{CBO} = \alpha\, I_E$$

$$\frac{I_C}{\alpha} - \frac{I_{BCO}}{\alpha} = I_E = I_C + I_B$$

$$\frac{I_C}{\alpha} - I_C = I_B + \frac{I_{CBO}}{\alpha}$$

$$I_C\left[\frac{1}{\alpha} - 1\right] I_B + \frac{I_{CBO}}{\alpha}$$

$$I_C = I_B\left[\frac{\alpha}{1-\alpha}\right] + \frac{I_{CBO}}{1-\alpha} \tag{4.8}$$

The ratio of the input current $I_B$ to the output current $I_C$ in a common emitter arrangement is known as the *current gain or Common Emitter amplification factor.*

$$\beta = \frac{I_C}{I_B} \tag{4.9}$$

We have, $I_E = I_C + I_B$ i.e., $I_B = I_E - I_C$

$$\beta = \frac{I_C}{I_E - I_C} \tag{4.10}$$

By dividing the numerator and the denominator of the equation's R.H.S by IE, we obtain,

$$\beta = \frac{\frac{I_C}{I_E}}{\frac{I_E}{I_E} - \frac{I_C}{I_E}}$$

$$\beta = \frac{\alpha}{1-\alpha} \tag{4.11}$$

We know that, $\alpha = \frac{I_C}{I_E}$ and $I_E = I_B + I_C$

$$\alpha = \frac{I_C}{I_B + I_C} \tag{4.12}$$

When the numerator and the denominator are divided by IB, we obtain,

$$\alpha = \frac{\frac{I_C}{I_B}}{\frac{I_B}{I_B} + \frac{I_C}{I_B}}$$

Then the value of $\alpha$ is,

$\alpha = \beta / (1 + \beta)$ (4.13)

## Input Characteristics

It represents the relationship between input voltage VBE and input current IB at a fixed output voltage VCE shown in Fig. 4.9. From the input properties of the common emitter arrangement, the input resistance is computed. The base-emitter voltage change ($\Delta V_{BE}$) to the consequent change in base current ($\Delta I_B$) at constant emitter voltage (VCE) is measured as the input resistance. It is written as,

$$r_i = \frac{\Delta V_{BE}}{\Delta I_B} \Big| V_{CE=constant} \quad (4.14)$$

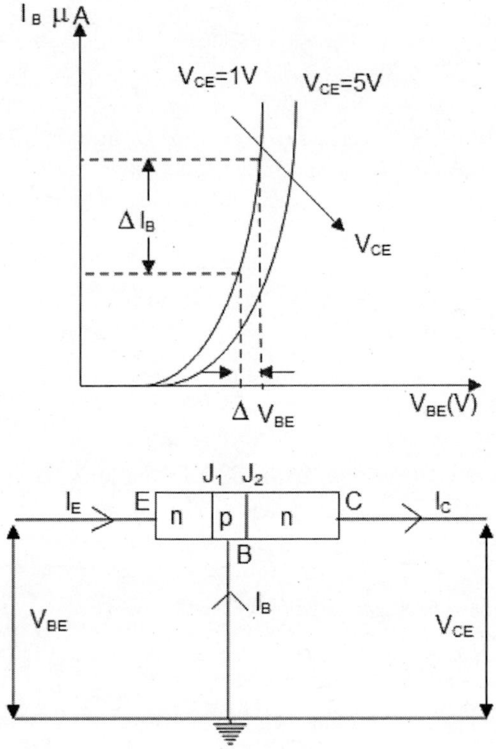


**Fig. 4.9:** Input Characteristics of CE configuration

The value of $r_i$ in CE configuration is greater compared to the CB configuration. For constant value of base to emitter voltage, the base current decreases as collector to emitter voltage is increased.

## Output Characteristics

It is the curve drawn between collector current $I_C$ and collector-emitter voltage $V_{CE}$ at constant base current $I_B$. The output characteristics of common emitter is shown in Fig. 4.10. keeping the base current $I_B$ fixed at some value, note the collector current $I_C$ varies for small range of $V_{CE}$. After this, collector current becomes constant and independent of $V_{CE}$. This value of $V_{CE}$ upto which collector current $I_C$

changes with $V_{CE}$ is called *Knee voltage.* Above knee voltage, $I_C$ is almost constant. For any value of $V_{CE}$ above knee voltage, the collector current approximately equal to $\beta \times I_B$. The output resistance is defined as the ratio of change in collector-emitter voltage to the change in collector current. In the CE arrangement, the output dynamic resistance is high.

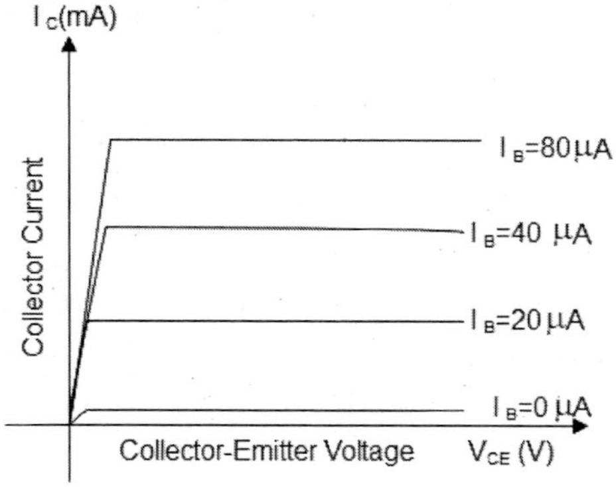


**Fig. 4.10:** Output Characteristics of CE Configuration.

$$r_o = \frac{\Delta V_{CE}}{\Delta I_C} \bigg| \; I_{B=constant} \tag{4.15}$$

**Example 4.3:** Calculate the collector and emitter current for a BJT with $\alpha = 0.99$ and $I_B = 20\mu A$.

**Given:** $\alpha = 0.99$, $I_B = 20\mu A$

**Solution:**

The current amplification factor β is given by,

$$\beta = \frac{\alpha}{1-\alpha} = \frac{0.99}{1-0.99} = 99$$

We know that $\frac{I_C}{I_B} = \beta$, the collector current calculated as,

$$I_c = \beta I_B = 99 \times 20 \times 10_{-6} = 1.98mA$$

The emitter current is the sum of both collector and base current is expressed as,

$$I_E = I_C + I_B = 1.98mA + 20\mu A = 2mA$$

**Example 4.4:** Find the values of $I_C$ and $I_E$ for a BJT with $\alpha = 0.97$ and $I_B = 50\mu A$. Determine β for the device.

**Given:** $\alpha = 0.97$, $I_B = 50\mu A$

**Solution:** Current amplification factor β is given by,

$$\beta = \frac{\alpha}{1-\alpha} = \frac{0.97}{1-0.97} = 32.33$$

We know that $\frac{I_C}{I_B} = \beta$, the collector current calculated as,

$$I_C = \beta I_B = 32.33 \times 50 \times 10^{-6} = 1.6165\text{mA}$$

The emitter current is the sum of both collector and base current is expressed as,

$$I_E = I_B + I_C = 50\mu\text{A} + 1.6165\text{mA}$$

$$= 1.6665\text{mA}$$

## 4.6.3 Common Collector Configuration

In a conventional collector design, input is applied between the base and the collector. The output is provided by the emitter and the collector terminals. This structure is referred as a common collector configuration or an emitter follower configuration as the transistor's collector is shared by the input and the output circuits. The collector configuration for both npn and pnp transistors are shown in Fig. 4.11.

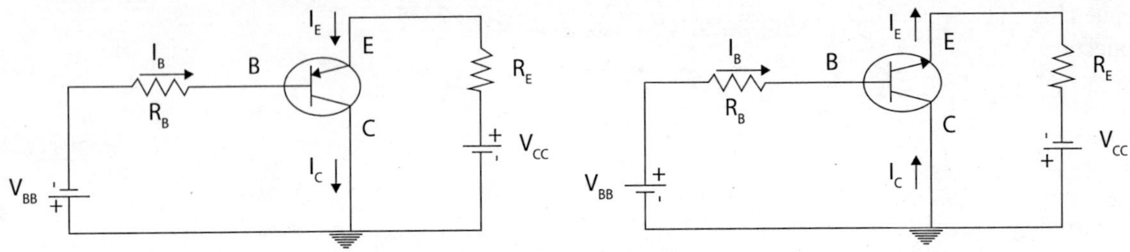


pnp transistor

npn transistor

**Fig. 4.11:** Common Collector Configuration

### Current Relation in CC Configuration

We know that,

$$I_E = I_B + I_C$$

$$= I_B + \alpha I_E + I_{CBO}$$

$$I_E (1 - \alpha) = I_B + I_{CBO}$$

$$I_E = \frac{I_B}{1 - \alpha} + \frac{I_{CBO}}{1 - \alpha}$$

Scan to know more

We know that,

$$\beta = \frac{\alpha}{1 - a}$$

$$1 + \beta = 1 + \frac{\alpha}{1-\alpha} = \frac{1-\alpha+\alpha}{1-\alpha} = \frac{1}{1-\alpha}$$

$$I_E = I_B\,(1 + \beta) + I_{CBO}\,(1 + \beta)$$

Neglecting $I_{CBO}$ we have,

$$I_E = I_B\,(1 + \beta) \tag{4.16}$$

The current gain in CC configuration is given by,

$$\gamma = \frac{I_E}{I_B} = 1 + \beta \tag{4.17}$$

### *Input Characteristics*

The input characteristics is drawn between the input current $I_B$ and the output voltage $V_{CB}$ for various values of collector to emitter voltage $V_{CE}$.

Common base and common emitter input characteristics are very different from common collector input characteristics. This discrepancy results from the input voltage being mostly dictated by the collector to emitter voltage.

From the Fig. 4.12 we can write,

$$V_{CE} = V_{CB} + V_{CE}$$

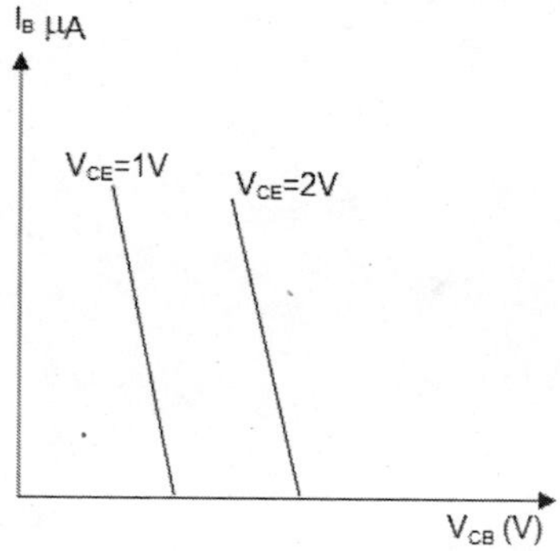


**Fig. 4.12:** Input Characteristics of CC Configuration

$$V_{CB} = V_{CE} - V_{BE}$$

In CC configuration the input junction is base to collector, and it is reverse biased. So that the input resistance in CC configuration is very high.

### *Output Characteristics*

Output characteristics are drawn between the output current ($I_E$) and the output voltage ($V_{CE}$) for varied base current ($I_B$).

The common collector output characteristics are virtually identical to those of the common emitter output configuration since the collector current is almost equal to the emitter current.

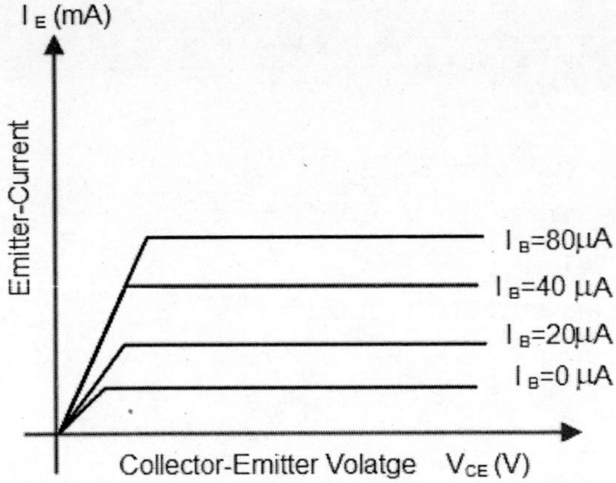


**Fig. 4.13:** Output Characteristics of CC Configuration

## 4.7 COMPARATIVE ANALYSIS OF CB, CE AND CC CONFIGURATIONS

**Table 4.2:** Comparison of CB, CE, CC configurations

| *Input Characteristics* | *CB Configuration* | *CE Configuration* | *CC Configuration* |
|---|---|---|---|
| Resistance at input | Very low | Low | High |
| Resistance at output | Very high | High | Low |
| Input current | $I_E$ | $I_B$ | $I_B$ |
| Output current | $I_C$ | $I_C$ | $I_E$ |
| Leakage current | Small | Large | Large |
| Amplification factor | $\alpha = I_C / I_E$ | $\beta = I_C / I_B$ | $\gamma = I_E / I_B$ |
| Current gain | Less than unity | High | High |
| Voltage gain | Medium | Medium | Less than unity |

## 4.8 BIASING OF TRANSISTOR

The main objective of a transistor is to amplify the input signal. However, there should be no change in the signal space; only the signal's magnitude should grow. We refer to this as *faithful amplification.*

### 4.8.1 Need for Biasing

The base-emitter junction of the transistor must always be forward biased to do amplification. Reverse bias is always present at the collector base configuration. Transistor biasing is the term used for this.

### 4.8.2 Operating Point

The transistor node voltages and currents can be considered as a combination of two terms–DC quantity and variation around the DC quantity. This DC quantity is known as operating point or *Quiescent point*

*(Q-point)* and variation around the DC quantity is known as AC or small signal. The zero signal values of $I_C$ and $V_{CE}$ in a transistor are known as the operating point for the proper operation of transistor as an amplifier. It is necessary to set up fixed d.c level of current through the transistor with fixed d.c voltage across it.

### 4.8.3 DC Load Line

In the transistor circuit analysis, it is generally required to determine the collector current for various collector-emitter voltages. For this analysis DC load line method is analysed. Any changes in the load ($R_C$), the value of $V_{CE}$ and $I_C$ points' locations are found using the DC Load-Line. To create a d.c. load line, we require two end points.

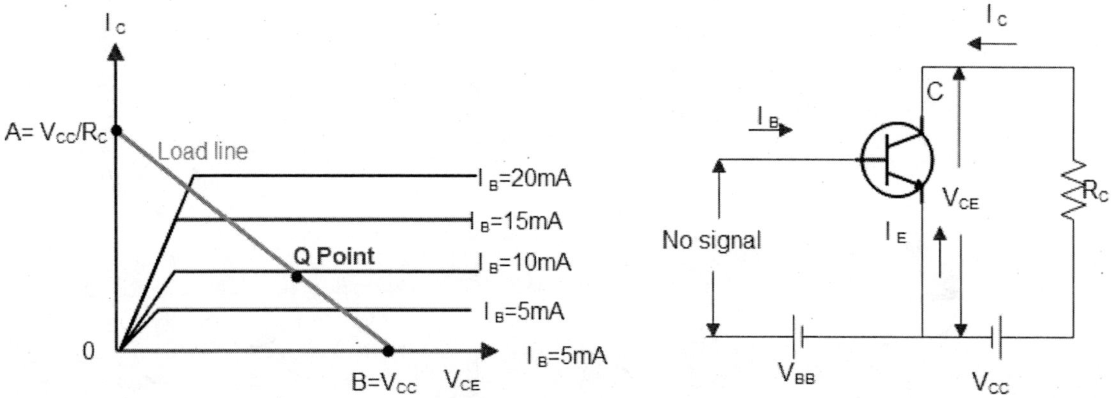


**Fig. 4.14:** Quiescent Point

i. **$I_c = 0$, $V_{CE}$ = Maximum**

   The collector-emitter voltage is high and equal to $V_{CC}$.When the collector current $I_C = 0$,

   I_C=0,

   Max $V_{CE} = V_{CC} - I_C R_C$

   Max $V_{CE} = V_{CC} - 0.R_C = V_{CC}$

   This results in the first collector emitter point B (OB = $V_{CC}$) shown in Fig. 4.14.

ii. **$V_{CE} = 0$, $I_C$ = Maximum**

iii. The maximum collector current, which is equal to $V_{CC} / R_C$ occurs at collector-emitter voltage $V_{CE} = 0$.

   This results in the second collector current point A (OA = $V_{CC}/ R_C$). These two points are connected to create DC line AB.

## 4.9 BIASING CIRCUIT

The biasing circuits are used to provide biasing and stabilization. The *voltage divider bias* method is the most prominent one. Here, two resistors $R_1$ and $R_2$ are employed, which are connected to $V_{CC}$ and provide biasing. The resistor $R_E$ employed in the emitter provides stabilization.

### 4.9.1 Circuit Diagram

The name voltage divider comes from the voltage divider formed by $R_1$ and $R_2$. The voltage drop across $R_2$ forward biases the base-emitter junction. This causes the base current and hence collector current flow in the zero signal conditions. The Fig. 4.15 shows the circuit diagram of voltage divider bias method. It is also called as *self-biasing circuit.*

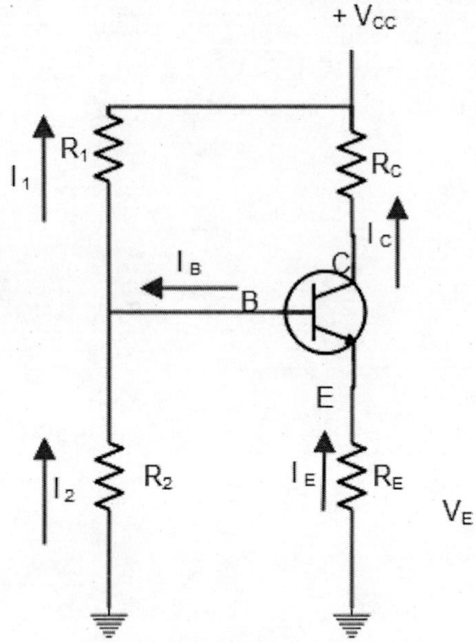


**Fig. 4.15:** Voltage Divider Bias Circuit

Assume that $I_1$ is the current flowing through resistor $R_1$. It can be reasonably believed that current flowing through $R_2$ is also $I_1$ because base current $I_B$ is relatively very small.

The operating points ($I_C$,$V_E$) will be found using Kirchhoff's voltage law.

From the circuit, the current $I_1$ is found by,

$$I_1 = \frac{V_{CC}}{R_1 + R_2} \tag{4.18}$$

Voltage across the resistor $R_2$,

$$V_2 = \left[\frac{V_{CC}}{R_1 + R_2}\right] R_2 \tag{4.19}$$

Apply KVL to the base circuit,

$$V2 = V_{BE} + V_E$$

or

$$V_2 = V_{BE} + I_E R_E$$

The emitter current, $I_E$ is,

$$I_E = \frac{V_2 - V_{BE}}{R_E}$$

Since, $I_E = I_C$

$$I_C = \frac{V_2 - V_{BE}}{R_E} \tag{4.20}$$

It is clear from the equation 4.20 that $I_C$ does not at all depend upon β. $V_{BE}$ is very small such that IC does not affect by $V_{BE}$. So, $I_C$ in the circuit almost independent of transistor parameters and hence good stabilization is ensured.

To find the collector-emitter voltage $V_{CE}$,

Apply KVL to the collector circuit,

$$V_{CC} = I_C R_C + V_{CE} + I_E R_E$$

$$= I_C R_C + V_{CE} + I_C R_E$$

$$= I_C (R_C + R_E) + V_{CE}$$

$$V_{CE} = V_{CC} - I_C (R_C + R_E)$$

The voltage drop across resistor $R_2$ is given by,

$$V_2 = V_{BE} + I_C R_E$$

The resistance $R_E$ provides the good stabilization. If the temperature rises, the collector current $I_C$ drops, causing the voltage drop across $R_E$ to increase. The value of $V_{BE}$ drops when the voltage drop across $R_2$ is $V_2$, which is independent of the $I_C$. The lower $I_B$ value tends to return $I_C$ to its original value. The operating point of voltage divider bias is drawn between ($I_C$, $V_{CE}$). The main advantage of voltage divider bias is that stability is high compared to another biasing scheme.

### 4.9.2 Stability Factor

The stability factor of a voltage divider bias refers to a measure of how stable the biasing conditions are for the transistor. The stability factor is typically represented by the symbol "S" and is calculated as the ratio of the change in the base-emitter voltage ($V_{BE}$) to the change in the collector current ($I_C$) when the transistor parameters vary. Stability factor (S) of the circuit is given by,

$$S = \frac{(\beta + 1)(R_T + R_E)}{R_T + R_E + \beta R_E} \tag{4.21}$$

$$= (\beta + 1) \times \frac{1 + \frac{R_T}{R_E}}{\beta + 1 + \frac{R_T}{R_E}} \tag{4.22}$$

where, $R_T = \dfrac{R_1R_2}{R_1 + R_2}$. If the ratio $R_T/R_E$ is very small, then $R_T/R_E$ can be neglected as compared to 1 and the stability factor becomes,

$$\text{Stability factor } (S) = (\beta + 1) \times \frac{1}{(\beta + 1)} = 1 \tag{4.23}$$

This is the smallest possible value of S and leads to the maximum possible thermal stability. Increasing the value of $R_E$ requires greater $V_{CC}$ to maintain the same value of zero signal collector current. Therefore, the ratio $R_T/R_E$ cannot be made very from design point of view.

**Example 4.5:** Consider the figure 4.15, the value $R_1 = 50K\Omega$, $R_2 = 10K\Omega$, $R_C = 2K\Omega$ and $R_E = 1K\Omega$. If $V_{CC} = 12V$. Find the value of $I_C$, given $V_{BE} = 0.1V$

**Given:** $R_1 = 50K\Omega$, $R_2 = 10K\Omega$, $R_C = 2K\Omega$, $R_E = 1K\Omega$ and $V_{CC} = 12V$.

**Solution:**

$V_{BE} = 0.1V$

Collector current, $I_C = \dfrac{V_2 - V_{BE}}{R_E}$,

The voltage across $R_2$,

$$V_2 = \frac{R_2}{R_1 + R_2} V_{CC} = \frac{10}{50 + 10} \times 12 = 2V$$

The current, $I_C = \dfrac{V_2 - V_{BE}}{R_E} = \dfrac{2 - 0.1}{1K\Omega} = 1.9mA$

## 4.10 EBERS-MOLL MODEL

In order to analyse transistor circuits, equivalent circuit models are developed. The Ebers-Moll model, or equivalent circuit, is one of the classic models of the bipolar junction transistor. It is used to analyse the switching application of the transistor. Switching usually involves turning a transistor from its "off" state, or "cutoff", to its "on" state. The model has two current sources $I_F$, $I_R$ and two diodes, $D_1$ and $D_2$. The base-emitter and base-collector junctions are represented by diodes. Fig. 4.16 shows the current directions and the voltage polarities used in the Ebers-Moll model.

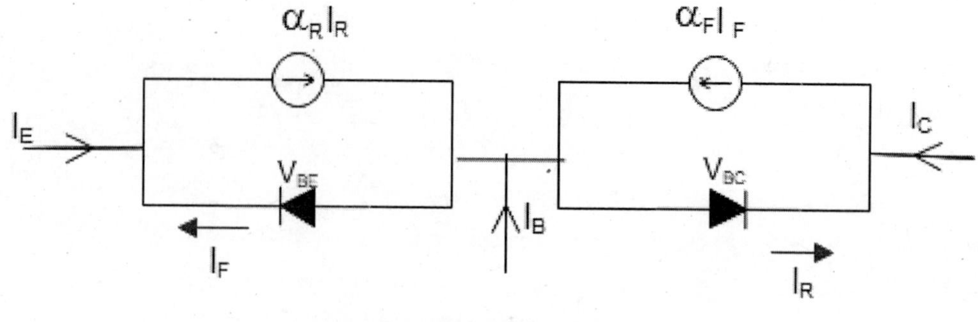


**Fig. 4.16:** Ebers-Moll Model

The sum of currents entering the terminal is shown as,

$$I_E + I_B + I_C = 0$$

The collector current can be written in general as,

$$I_c = -I_R + \alpha_F I_F \tag{4.24}$$

where, $\alpha_F$ is the common base current gain in the forward active mode. In this mode of operation, the collector current is defining as,

$$I_c = -I_{CO} + \alpha_F I_F$$

where the current $I_{CO}$ is the reverse bias B-C junction current. The forward current $I_F$ is given by,

$$I_F = I_{EO}\left(\frac{V_{BE}}{e^{\eta VT}} - 1\right) \tag{4.25}$$

If the B-C junction becomes forward biased, such as in saturation, then the reverse current $I_R$ is written as,

$$I_R = I_{CO}\left(e^{\frac{V_{BC}}{\eta VT}} - 1\right) \tag{4.26}$$

$I_{EO}$ is the reverse bias B-E junction current. We also write the emitter current as,

$$I_E = I_F - \alpha_R I_R \tag{4.27}$$

where, $\alpha_R$ is the common base current gain for inverse active mode.

$$I_B = I_E - I_C \tag{4.28}$$

$$= -\alpha_R I_R + I_F + I_R - \alpha_F I_F \tag{4.29}$$

$$= I_F(1 - \alpha_F) + (1 - \alpha_R) I_R \tag{4.30}$$

The Ebers-Moll Model has four parameters $\alpha_F$, $\alpha_R$, $I_{EO}$ and $I_{CO}$. These parameters are related by,

$$I_{EO}\,\alpha_F = I_{CO}\,\alpha_R \tag{4.31}$$

By analysing the minority carrier current passing through the base, one can deduce the relationship, this is also known as the reciprocity relation.

From equation 4.27 we have,

$$I_E = I_F - \alpha_R I_R$$

Substitute $I_R = \alpha_F I_F - I_C$ from equation 4.24,

$$= I_F - \alpha_R(\alpha_F I_F - I_C)$$

$$= I_F - \alpha_R\,\alpha_F I_F + \alpha_R\,I_C$$

$$= \alpha_R\,I_C + (1 - \alpha_R\alpha_F)\,I_F$$

$$= \alpha_F I_E - (1 - \alpha_F\alpha_R)\,I_{CO}\left(e^{\frac{V_{BC}}{\eta VT}} - 1\right) \tag{4.32}$$

Scan to know more

From equation 4.32 the base emitter voltage of a transistor can be obtained as below,

$$e^{\frac{V_{BC}}{\eta VT}} - 1 = \frac{1E - \alpha_R I_C}{(1-\alpha_R \alpha_F) I_{EO}}$$

$$V_{BE} = \eta V_T \ln\left[1 + \frac{I_E - \alpha_R I_C}{(1-\alpha_R \alpha_F)\, I_{EO}}\right] \tag{4.33}$$

The value of $V_{BE}$ at cut-off can be obtained by substituting $I_E = 0$ and $I_c = -I_{CO}$ in the above equation.

$$V_{BE,cutoff} = \eta V_T \ln\left[1 + \frac{\alpha_R I_{CO}}{(1-\alpha_R \alpha_F)\, I_{EO}}\right] \tag{4.34}$$

$$\eta V_T \ln\left[1 + \frac{\alpha_F}{(1-\alpha_R \alpha_F)}\right] \tag{4.35}$$

## UNIT SUMMARY

- For npn BJT, base width << Ln i.e. diffusion length of electrons
- For pnp BJT, base width << Lp i.e. diffusion length of holes
- Modes of operation

**Active:** used significantly in analog circuits

Emitter to Base–Forward bias

Collector to Base–Reverse bias

**Saturation:** (On state in digital)

Emitter to Base–Forward bias

Collector to Base–Forward bias

**Cut-off:** (Off state in digital)

Emitter to Base–Reverse bias

Collector to Base–Reverse bias

### Reverse active:

Emitter to Base–Forward bias

Collector to Base–Reverse bias

- Base width should be minimum as possible to improve the current
- Small base width provides more stability
- Stability should be maintained to obtain high efficiency
- Some notations:

$$\alpha = \frac{I_C}{I_E}$$

Base transport factor, $\propto_T = I_{PC} / I_{PE}$

Emitter injection efficiency, $\gamma = I_{PE} / I_E$

$$\beta = \frac{I_C}{I_E} = \alpha/(1 - \alpha)$$

- $V_{EB} > 0$ = Forward bias–high diffusion
- $V_{CB} < 0$ = Reverse bias

## EXERCISES

### Multiple Choice Questions

4.1 The transistor operation is carried out by two types of charge carriers (majority and minority carriers), then transistor is called as____________

a. BJT b. MOSFET
c. JFET d. FET

4.2 In a junction transistor, the collector cut-off current $I_{CBO}$ reduces considerably by doping the

a. Emitter with a high level of impurity b. Emitter with a low level of impurity
c. Collector with a high level of impurity a. Collector with a low level of impurity

4.3 A transistor's current gain of 0.99 in the CB mode. Its current gain in the CC mode is

a. 100 b. 99
c. 1.01 d. 0.99

4.4 The Bipolar Transistor basic construction consists of two PN – junctions producing________ connecting terminals

a. Zero b. One
c. Two d. Three

4.5 BJT consists of ________ differently doped semiconductor regions

a. Three b. Two
c. One d. Zero

4.6 ____________ are constructed by stacking three layers of semiconductor material

a. Oscillators b. Transistors
c. Amplifiers d. Diodes

4.7 The primary function of BJT is to __________ the strength of a weak signal

a. Linear b. Decrease
c. Increase d. None

4.8 The ____________ is an electronic depiction of a transistor in one of the four basic configurations, either NPN or PNP

a. All b. High signal model
c. Small signal model d. Ebers-Moll Model

4.9 Fundamental assumption of the Ebers-Moll model that __________ fault occur in circuit

a. Only one b. Only two
c. Both a & b d. None

4.10 The main reason CE setup is utilised is __________

a. Low voltage gain
b. High current gain
c. Both a & b
d. None

4.11 Due to the substantially higher resistance in the common emitter circuit, the CE circuit is chosen over the CB circuit in amplifiers __________ than that of the common–base circuit

a. High
b. Medium
c. Less
d. None

4.12 Which transistor configuration is ideal (best)?

a. CB
b. CC
c. Both a & b
d. CE

4.13 The transistor's base terminal will be wired to the input and output in series, the configuration is __________

a. CB
b. CC
c. CE
d. None

4.14 The arrangement of the emitter between the collector and base is referred to as __________

a. CB
b. CE
c. CC
d. None

4.15 The arrangement where the collector is shared by the emitter and base is referred to as __________

a. CB
b. CE
c. CC
d. None

4.16 When no input current is applied to the component, the transistor is in the following state __________

a. R-point
b. L-point
c. I-point
d. Q-point

4.17 The location of the operational point is crucial for achieving faithful __________

a. Amplification
b. Oscillation
c. Both a & b
d. None

4.18 The biasing problem is that of establishing a constant ______ current at the collector of the BJT

a. AC
b. DC
c. Both AC and DC
d. None

4.19 BJT amplifiers does not ________ the necessary output across the load terminals in the absence of transistor biasing

a. Receive
b. Recover
c. Deliver
d. None

4.20 __________ identifies the DC operating point necessary for an amplifier to operate linearly

a. Switch
b. Amplify
c. Transistor
d. Bias

4.21 ____________ configuration can offer good bias stability

a. Voltage divider bias
b. Current divider bias
c. Both a & b
d. None

4.22 In a junction transistor, recombination of electrons and holes occurs in

a. Base region only
b. Emitter region only
c. Collector region only
d. All the three regions

| Answer of Multiple Choice Questions | | | | |
|---|---|---|---|---|
| 4.1 **a** | 4.2 **a** | 4.3 **a** | 4.4 **d** | 4.5 **a** |
| 4.6 **b** | 4.7 **c** | 4.8 **d** | 4.9 **a** | 4.10 **b** |
| 4.11 **c** | 4.12 **d** | 4.13 **a** | 4.14 **b** | 4.15 **c** |
| 4.16 **d** | 4.17 **a** | 4.18 **b** | 4.19 **c** | 4.20 **d** |
| 4.21 **a** | 4.22 **d** | | | |

## SHORT AND LONG ANSWER TYPE QUESTIONS

### Category I

4.1 Why is the device aptly named transistor?

4.2 Describe the basic structure of BJT.

4.3 What are the three terminals in a BJT?

4.4 What base-emitter and base-collector junction biases are required for a transistor to function as an amplifier?

4.5 Define Beta.

4.6 Why are silicon chosen over germanium while making semiconductor devices?

4.7 Define the relationship between $\alpha$ and $\beta$.

4.8 Which configuration has the lowest input impedance?

4.9 Define operating point of a transistor.

4.10 What is stability.

4.11 Describe the need for biasing.

4.12 What is Biasing?

4.13 Draw the circuit for npn and pnp transistor.

### Category II

4.1 Explain in detail about BJT construction.

4.2 Describe the working principle of BJT with suitable diagram.

4.3 Compare CB, CE, and CC configuration.

4.4 Illustrate CB configuration with neat sketch.

4.5 Explain CE configuration.

4.6 Explain in detail about CC configuration.

4.7 Draw and explain Ebers-Moll model of a transistor.

4.8 Define early effect.

4.9 Describe how the amplification and switching are achieved by a BJT.

4.10 Explain about DC load line.

## Numerical Problems

4.1 If the collector current is 2mA and the base current is 25μA, what is the emitter current?
**(Ans: 2.025mA)**

4.2 If $\alpha = 0.98$ what is the value of $\beta$.
**(Ans: 49)**

4.3 An amplifier has an input voltage of 0.01V. If the output is 40V, what is the value of voltage gain?
**(Ans:4000)**

4.4 Common base connection has the base current of 0.05mA and the collector current 1.95mA. Find the emitter current.
**(Ans: 2mA)**

4.5 Calculate the current at base of common-base configuration whose $\alpha$ is 0.98, emitter current 2mA.
**(Ans: 1.02mA)**

4.6 In common-base configuration, the current at the emitter is 2mA, $I_{CBO} = 10\mu A$, $\alpha = 0.92$. Find the total collector current.
**(Ans: 1.85mA)**

4.7 The change in collector current is 1.4mA. For a change in base current of 40μA, Find the value of $\beta$.
**(Ans: 35)**

4.8 In a common base configuration. Find base current for the given $I_E = 2mA$, IC = 1.9mA.
**(Ans:0.01mA)**

4.9 Find the value of $\beta$. If a) $\alpha = 0.98$, b) $\alpha = 0.99$, c) $\alpha = 0.1$.
**(Ans: a) 49, b) 99, c) 0.11)**

4.10 In a common base configuration, the value of $I_C = 0.96mA$ and base current is 0.02mA. Calculate the value of $\alpha$.
**(Ans: 0.97)**

4.11 In a common-base configuration, the emitter is 1mA. Consider the emitter side is open and the collector current is 10μA. Calculate the total collector current. If $\alpha = 0.93$.
**(Ans: 0.94mA)**

4.12 Calculate emitter current in a transistor whose $\beta = 50$, $I_B = 25\mu A$.
**(Ans: 1.275mA)**

## PRACTICAL

**Input and output characteristics of BJT in CB, CE, and CC configurations**

### *Aim*

To study the Characteristic of Common-Collector Configuration, common base, and common emitter configurations and to find input impedance of each configuration.

1. **Common Emitter Configuration**

## Apparatus Required

- Power supply–(0–30V)
- Ammeter (0–10mA), (0–1mA)
- Voltmeter (0–30V), (0–2V)

## Circuit Diagram

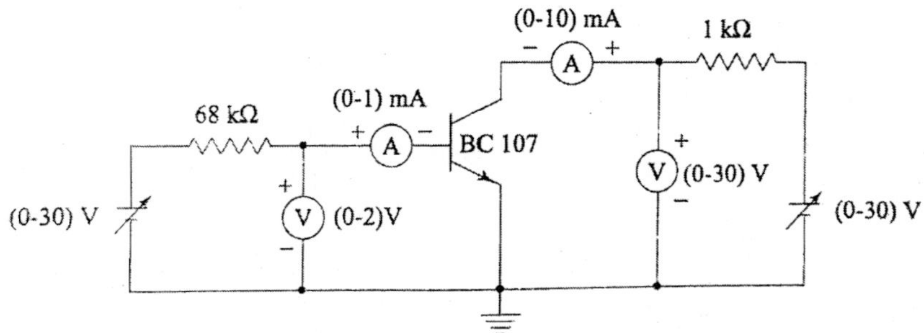


## Procedure

i. **Input Characteristics**

1. Assemble the circuit in accordance with the schematic.
2. Set $V_{CE}$ to 5V (for example), adjust $V_{BE}$ in 0.1V increments and record the associated $I_B$, follow the same steps for 10 V, I5 V, etc.
3. Draw the graph: $V_{BE}$ VS $I_B$ for a fixed $V_{CE}$.
4. Determine h-parameters: a. $h_{fe}$- forward current gain, b. $h_{ie}$- input impedance

ii. **Output Characteristics**

1. Assemble the circuit in accordance with the schematic.
2. Set IB to 20μA, adjust $V_{CE}$ in 1V increments and record the associated $I_C$, follow the same steps for 40μA, 80μA, etc.
3. Draw the graph: $V_{CE}$ vs $I_C$ for a fixed $I_B$.
4. Determine h-parameters: a. $h_{oe}$- forward current gain, b. $h_{re}$- input impedance

## Sample Graphs

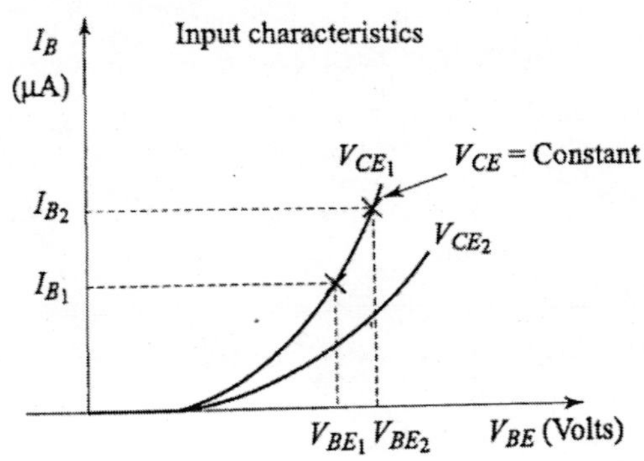


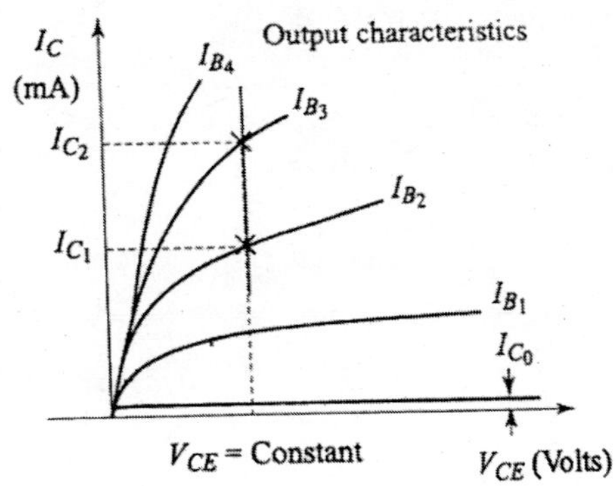

$$h_{ie} = \frac{\Delta V_{BE}}{\Delta I_B}\bigg|V_{C_{E=constant}} = \frac{V_{BE1} - V_{BE1}}{I_{B2} - I_{B1}}$$

$$h_{re} = \frac{\Delta V_{BE}}{\Delta I_{CE}}\bigg|V_{I_{B=constant}} = \frac{V_{BE2} - V_{BE1}}{V_{CE2} - V_{CE1}}$$

$$h_{fe} = \frac{\Delta I_C}{\Delta I_B}\bigg|_{V_{CE=constant}} = \frac{I_{CE2} - I_{C1}}{I_{B2} - I_{B1}}$$

$$h_{re} = \frac{\Delta I_C}{\Delta I_{CE}}\bigg|_{V_{IR=constant}} = \frac{I_{C2} - I_{C1}}{I_{CE2} - I_{CE1}}$$

## Tabulation

Characteristics of Input: $V_{CE}$ constant

| $V_{BE}$ (V) | $I_B$ ($\mu$A) |
|---|---|
| | |

Characteristics of Output: $I_B$ constant

| $V_{CE}$ (V) | $I_C$ (mA) |
|---|---|
| | |

## Result

| Parameters | Practical Values |
|---|---|
| $h_{fe}$ | |
| $h_{ie}$ | |
| $h_{re}$ | |
| $h_{oe}$ | |

2. **Common Collector Configuration**

## Apparatus Required

- Power supply–(0–30V)
- Ammeter (0–10mA), (0–1mA)
- Voltmeter (0–30V), (0–2V)

## Circuit Diagram

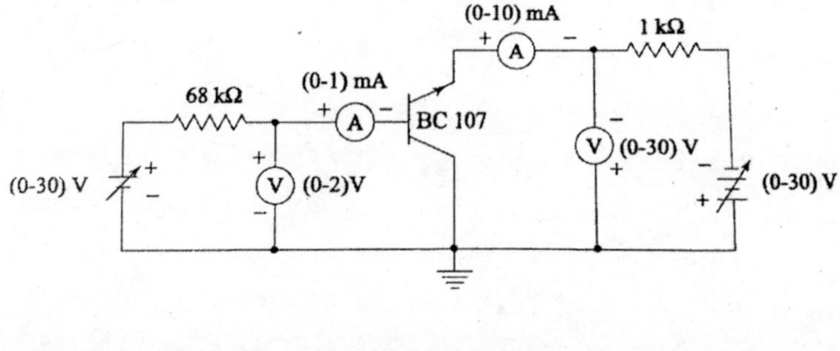

## *Procedure*

i. **Input Characteristics**

1. Assemble the circuit in accordance with the schematic.
2. Set $V_{EC}$ to 5V (for example), adjust $V_{BC}$ in of 0.1V increments and record the associated $I_B$, follow the same step for 10 V, I5 V, etc.
3. Draw the graph of $V_{BC}$ vs $I_B$ for a fixed $V_{EC}$.
4. Determine h-parameters: a. $h_{fc}$- forward current gain,
   b. $h_{ic}$- input impedance

ii. **Output Characteristics**

5. Assemble the circuit in accordance with the schematic.
6. Set $I_B$ to 20μA , adjust $V_{EC}$ in 1V increments, and record the associated $I_C$, follow the same steps for 40μA, 80μA, etc.
7. Draw the graph: $V_{CE}$ vs $I_C$ for a fixed $I_B$.
8. Determine h-parameters: a. hoc- forward current gain
   b. hrc- input impedance

## *Sample Graphs*

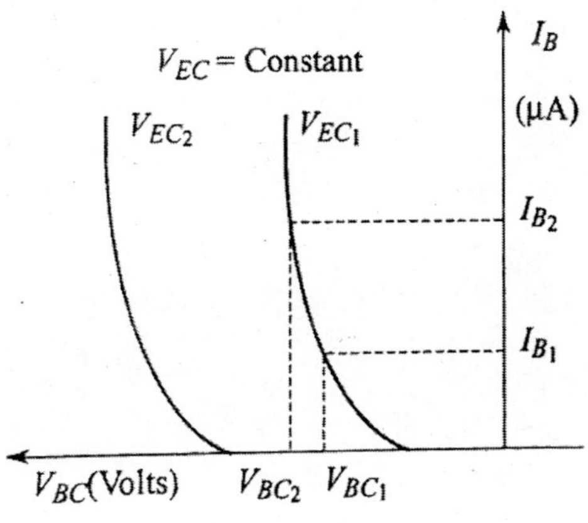


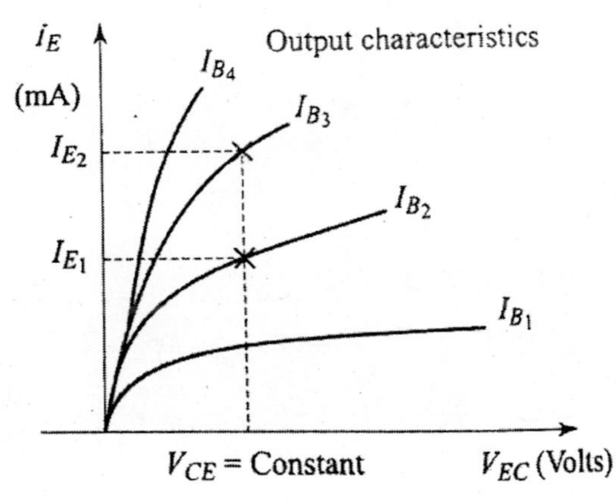


$$h_{ic} = \frac{\Delta V_{BC}}{\Delta I_B}\bigg|_{V_{CE=constant}} = \frac{V_{BC2} - V_{BC1}}{I_{B2} - I_{B1}}$$

$$h_{rc} = \frac{\Delta V_{BC}}{\Delta V_{CE}}\bigg|_{VB=constant} = \frac{V_{BC2} - V_{BC1}}{V_{CE2} - V_{CC1}}$$

$$h_{fc} = \frac{\Delta I_E}{\Delta I_B}\bigg|_{V_{EC=constant}} = \frac{I_{E2} - I_{E1}}{I_{B2} - I_{B1}}$$

$$h_{re} = \frac{\Delta I_E}{\Delta I_{EC}}\bigg|_{IB=constant} = \frac{I_{E2} - I_{E1}}{I_{EC2} - I_{EC1}}$$

## Tabulation

Characteristics of Input: $V_{EC}$ constant

| $V_{BC}$ (V) | $I_B$ ($\mu$A) |
|---|---|
| | |

Characteristics of Output: $I_B$ constant

| $V_{EC}$ (V) | $I_E$ (mA) |
|---|---|
| | |

## Result

| Parameters | Practical Values |
|---|---|
| $h_{fc}$ | |
| $h_{ic}$ | |
| $h_{rc}$ | |
| $h_{oc}$ | |

3. **Common Base Configuration**

## Apparatus Required

- Power supply–(0–30V)
- Ammeter (0–10mA), (0–1mA)
- Voltmeter (0–30V), (0–2V)

## Circuit Diagram

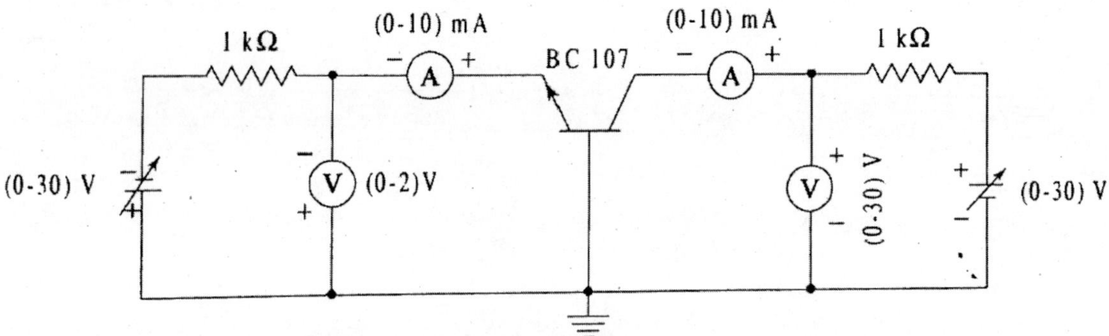


## Procedure

i. **Input Characteristics**

1. Assemble the circuit in accordance with the schematic.
2. Set $V_{CB}$ to 5V (for example), adjust $V_{EB}$ in 0.1V increments, and record the associated $I_E$, follow the same steps for 10 V, I5 V, etc.
3. Draw the graph of $V_{BE}$ vs $I_E$ for a fixed $V_{CB}$.

4. Determine h-parameters: a. $h_{fc}$- forward current gain
   b. $h_{ic}$- input impedance

ii. **Output Characteristics**
1. Assemble the circuit in accordance with the schematic.
2. Set $I_E$ to 20μA, adjust $V_{CB}$ in 1V increments, and record the associated $I_C$, follow the same steps for 40μA, 80μA, etc.
3. Draw the graph of $V_{CB}$ vs $I_C$ for a fixed $I_E$.
4. Determine h-parameters: a. $h_{oc}$- forward current gain
   b. $h_{rc}$- input impedance

## Sample Graphs

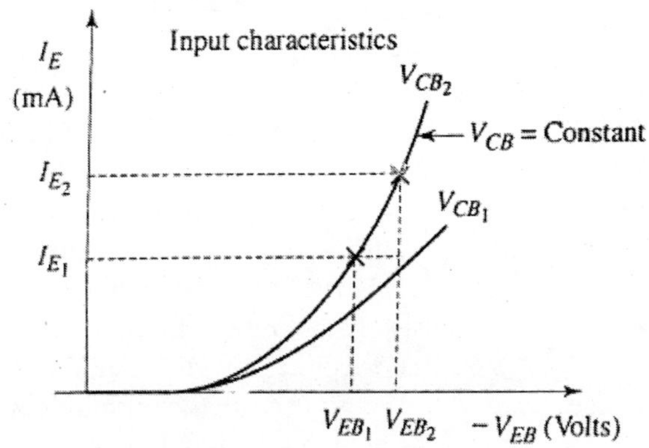


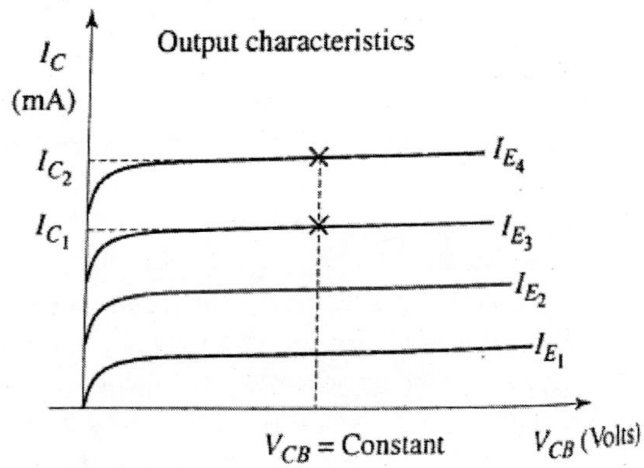


$$h_{ib} = \frac{\Delta V_{EB}}{\Delta I_E}\bigg|_{V_{CB=constant}} = \frac{V_{EB2} - V_{EB1}}{I_{E2} - I_{E1}}$$

$$h_{rb} = \frac{\Delta V_{EB}}{\Delta V_{CB}}\bigg|_{I_{E=constant}} = \frac{V_{EB2} - V_{EB1}}{V_{CB2} - V_{CB1}}$$

$$h_{fb} = \frac{\Delta I_C}{\Delta I_E}\bigg|_{V_{CB=constant}} = \frac{I_{C2} - I_{C1}}{I_{E2} - I_{E1}}$$

$$h_{re} = \frac{\Delta I_C}{\Delta I_{CB}}\bigg|_{I_{E=constant}} = \frac{V_{C2} - I_{C1}}{V_{CB2} - V_{CB1}}$$

## Tabulation

Characteristics of Input: $V_{CB}$ constant

| $V_{EB}$ (V) | $I_E$ (mA) |
|---|---|
| | |

Characteristics of Output: $I_E$ constant

| $V_{CB}$ (V) | $I_C$ (mA) |
|---|---|
| | |

### Result

| Parameters | Practical Values |
|---|---|
| $h_{fb}$ | |
| $h_{ib}$ | |
| $h_{rb}$ | |
| $h_{ob}$ | |

## STUDY EXPERIMENT

## Demonstrate Transistor as Switch

### Transistor Switching Time

When a transistor is in the active area, it may be employed as an amplifier. In this part, we will look at the use of a transistor as a switch. When used as a switch, a transistor is either in the ON or OFF state. When ON, the transistor is in the saturation area; when OFE, it is in the cut off zone. As a switch, the transistor has several applications. For example, a digital computer employs thousands of transistor switches. As a result, the speed with which a transistor acts as a switch are critical. Consider the temporal response of a basic transistor acting as a switch.The schematic of a transistor as a switch is shown below. The current applied to the base controls the switch. Figure depicts the transistor's reaction to input pulse V. This input pulse switches between voltage levels $V_a$ and $V_i$. The transistor is in cutoff mode at V and saturation mode at $V_a$. Even though the input pulse changes from $V_i$ to V at time t = 0, the output does not instantly respond to the input signal. Instead, there is a temporal lag between the input and output voltage transitions. Similarly, when the input waveform flips between a and $V_i$ at time T, the output does not instantly respond and needs some time to achieve the voltage V. We'll define some of the terms used in transistor switching times now.

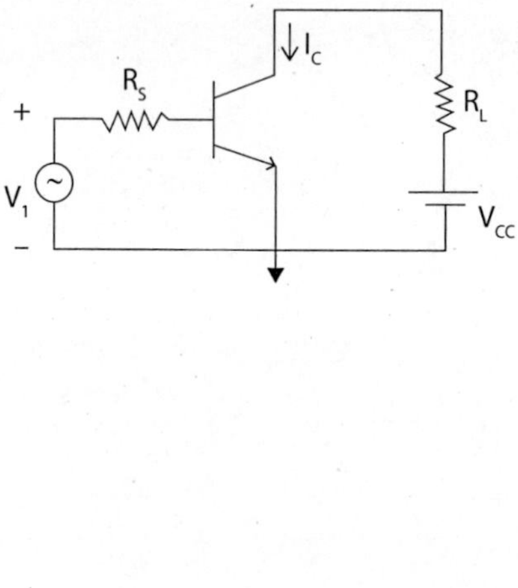


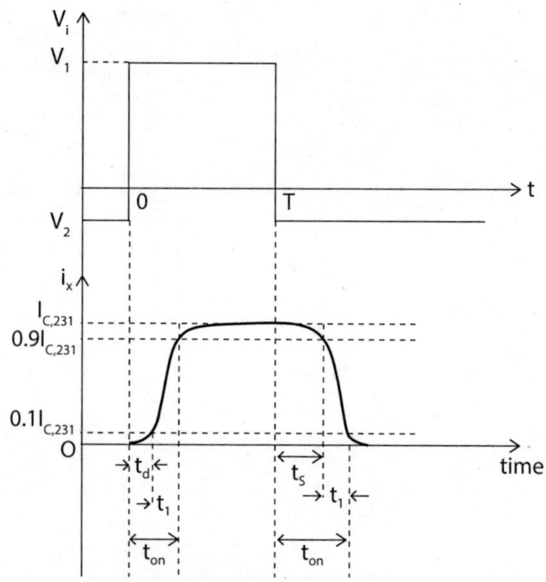

**Switching time-on**–The turn-on time is the amount of time it takes from the input voltage's leading edge to the collector current reaching 90% of its maximum value.

**Delay time**–It is the amount of time needed for the collector current to rise to 0.1 $I_{Csat}$. The delay time is affected by three things.

(i) When the transistor goes from off to on, it takes a non-zero amount of time to charge the emitter junction capacitance.

(ii) It takes a non-zero amount of time for minority carriers to pass the base region and subsequently reach the collector region.

Scan to know more

(iii) The collector current takes a while to increase to 10% of its maximum.

**Rise time**–It is the time necessary to boost the collector current from 0.1 $I_{Csat}$ to 0.9 $I_{Csat}$. During this time, the transistor is in the regular active zone. The rising time is caused by the transistor collector current having to transit the active area before reaching saturation from cut-off.

**Turn off time**–Turn-off time toff is the time necessary for the collector current to fall from $I_{Csat}$ to 0.1 $I_{Csat}$ when Vi turns negative.

**Storage time**–The storage time is the amount of time that elapsed between the input pulse's trailing edge and the time when it started to decline toward zero. This storage period is caused by the saturation surplus charge that is held in the transistor's base while it is in saturation. The transistor requires some time to eliminate the extra charge, which is referred to as storage time.

**Fall time**–The fall time is the length of time necessary for the collector current to decrease from $I_{Csat}$ to 0.1 $I_{Csat}$. The transistor is responsible for the fall time. Before reaching saturation cut-off, collector current must cross the active area.

**Propagation time**–The time it takes for a transistor switch to respond to an input signal is called propagation time Tpd.

## Switching Behaviour of Transistor

### *Aim*

To analyse the Switching behaviour of a Transistor

1. Rise time
2. Fall time
3. ON/OFF time
4. Delay time

### *Apparatus Required*

- Signal Generator–(0-1) MHz
- Cathode Ray Oscilloscope–(0-20) MHz
- Powersupply–(0-30) V

## Circuit Diagram

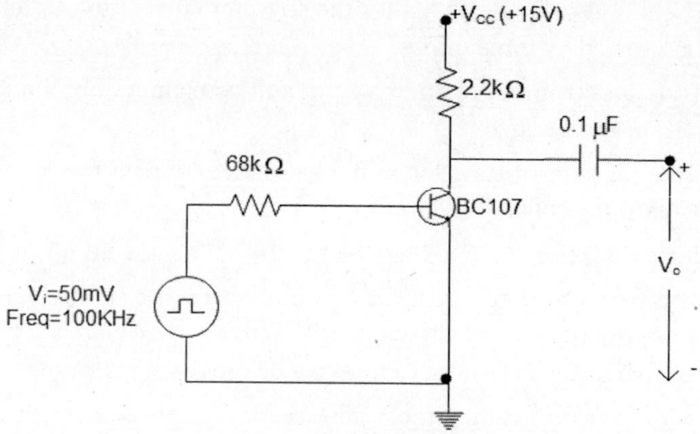


## Procedure

1. Rig up the circuit as per the circuit diagram.
2. Set input signal (say 50 mV, 100 kHz), using a signal generator.
3. Observe the output at the collector of the transistor using a CRO (AC-m0de).
4. Note down the parameters listed above and plot it on a graph.

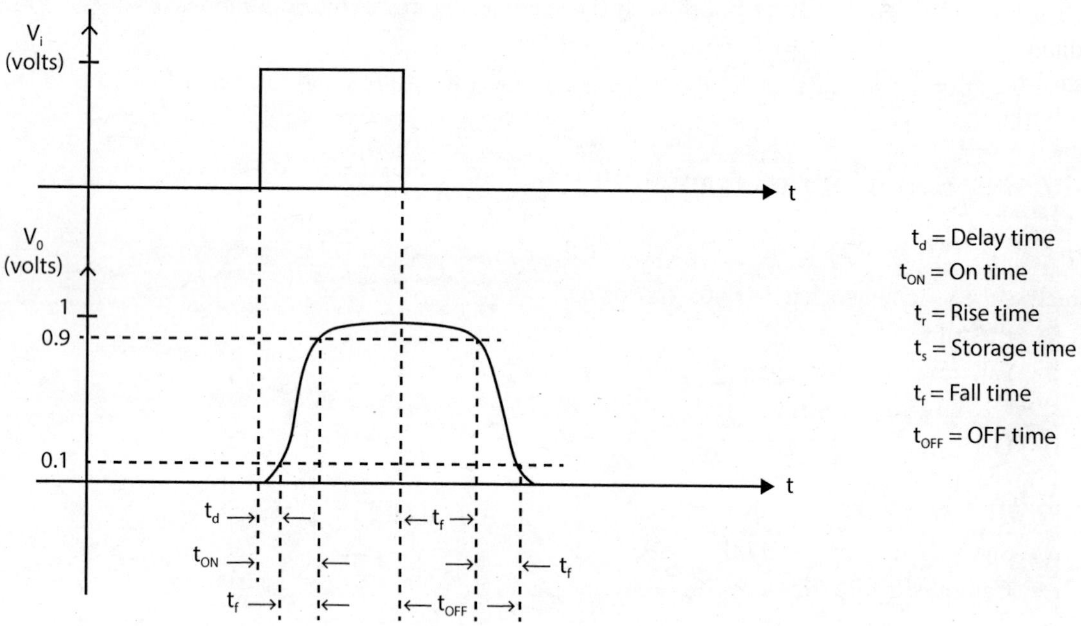

## Tabulation

| Parameters | Practical Values |
|---|---|
| Rise-time | |
| Fall-time | |
| On and off time | |
| Delay-time | |

## Result

Switching characteristics of the transistor is studied and their working is analysed.

## KNOW MORE

BJT is the most important topic in this chapter. The inventors of BJT received a Nobel prize in physics. At Bell Laboratories in the United States, a team led by William Shockley and included John Bardeen, Walter Brattain, and others invented the transistor in December 1947. Later, in 1951, he created the transistor's junction variant. In 1956, they were awarded the Nobel Prize for creating the transistor. Borden is the only physicist who received two times Nobel Prize in Physics. The first one was 1956 for the transistor. The second one was 1972 for BCS theory. The first transistor image is shown in figure. Shockley was irritated that the device was credited to Brattain and Bardeen, whom he believed had created it "behind his back" to steal the fame. When Bell Labs attorneys discovered that parts of Shockley's own writings on the transistor were similar enough to those of an earlier 1925 patent by Julius Edgar Lilienfeld that they deemed it prudent to leave his name off the patent application, things became worse.

Scan to know more

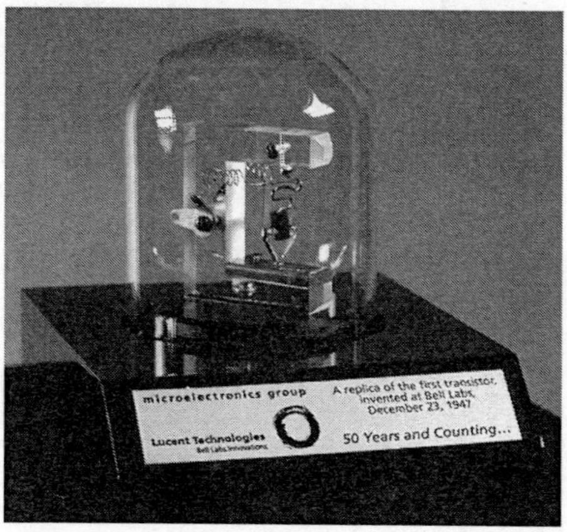


**First Transistor**

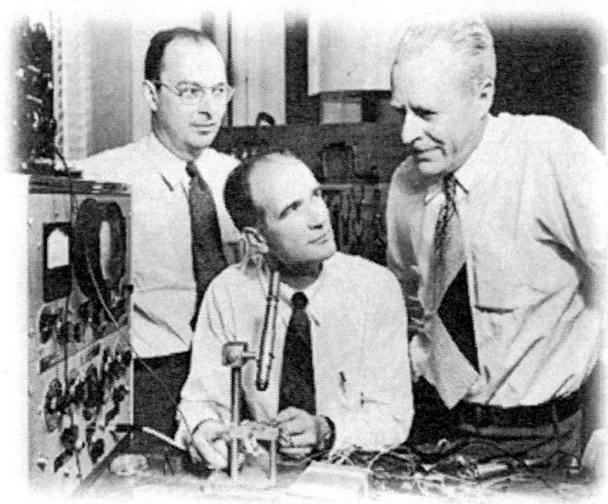

**Jhon Bardeen, Walter Brattain, William Shockly**

## REFERENCES AND SUGGESTED READINGS


1. David.J.Griffiths, *Introduction to Quantum Mechanics,* Pearson Prentice Hall, 2005.
2. Chenming Calvin Hu, *Modern Semiconductor Devices for Integrated Circuits,* Pearson Prentice Hall, 2009.
3. S.O.KASAP, *Principle of Electronic Materials & Devices,* MC Graw Hill Education, 2018.
4. S.Datta, *Quantum Transport:Atom to Transistor,* Cambridge University Press, 2005.
5. Robert F. Pierret, *Semiconductor Device Fundamentals,* ADDISON-WESLEY Publishing Company, 1996.
6. Donald A. Neamen, *Semiconductor Physics and Devices Basic Principle,* MC Graw Hill Companies, 2012.
7. Ben G, Streetman, Sanjay Kumar Banerjee, *Solid State Electronic Devices,* Pearson Education Limited, 2016.
8. *https://physicsinmyview.com/2020/09/facts-about-quantum-physics.html*
9. *https://onlinecourses.nptel.ac.in/noc21_ee80/course*
10. *https://onlinecourses.nptel.ac.in/noc21_ee59/course*
11. *https://onlinecourses.nptel.ac.in/noc21_ee19/course*
12. https://en.wikipedia.org/wiki/BJT.

# 05 FIELD EFFECT TRANSISTOR

## UNIT SPECIFICS

*Through this unit the following aspects are discussed:*

- *Construction and operation of Junction Field Effect Transistor*
- *Introduction of Metal-Oxide Semiconductor Field Effect Transistor*
- *Types of MOSFETs*
- *Transfer characteristics of MOSFET*
- *MOS C-V characteristics*
- *Explanation about MOS Capacitor*
- *Small Signal Model of MOS Transistor*

*The practical applications of the topics are discussed for generating further curiosity and creativity as well as improving problem solving capacity.*

*Besides giving many multiple-choice questions as well as questions of short and long answer types marked in two categories following the lower and higher order of Bloom's taxonomy, assignments through a number of numerical problems, a list of references and suggested readings are given in the unit so that one can go through them for practice. It is important to note that for getting more information on various topics of interest some QR codes have been provided in different sections that can be scanned for relevant supportive knowledge.*

*After the related practical, based on the content, there is a "Know More" section. This section has been carefully designed so that the supplementary information provided in this part becomes beneficial for the users of the book. This section mainly highlights the initial activity, examples of some interesting facts, analogious, history of the development of the subject focusing the salient observations and findings, timelines starting from the development of the concerned topics up to the recent time, applications of the subject matter for our day-to-day real life or/and industrial applications on a variety of aspects, case study related to environmental, sustainability, social and ethical issues whichever applicable, and finally inquisitiveness and curiosity topics of the unit.*

## RATIONALE

*This unit is on field effect transistor helps students to get a primary idea about the construction and operating characteristics of Junction Field Effect Transistor (JFET). It explains the basics of Metal-Oxide Semiconductor Field Effect Transistor and it extend to the operation and drain current characteristics of*

*MOSFET. All these basic aspects are relevant to study the advance semiconductor materials properly. It then clearly explains the structure of MOS capacitor as well as the operating conditions of MOS capacitor. Small signal modeling of MOSFET is elaborated to understand the operation in low frequency. All these are discussed at length to develop the subject. Some related problems are pointed out which can help further for getting a clear idea of the concern topics on field effect transistor.*

## PRE-REQUISITES

*Basics of Bipolar Junction Transistor*

## UNIT OUTCOMES

*List of outcomes of this unit is as follows:*

*U5-O1: Describe construction and working of Junction FET*

*U5-O2: Describe depletion and enhancement mode of MOSFET*

*U5-O3: Explain MOS Capacitor*

*U5-O4: Realize C-V Characteristics of MOS capacitor*

*U5-O5: Small signal Model of MOSFET*

| Unit-5 Outcomes | EXPECTED MAPPING WITH COURSE OUTCOMES (1-Weak Correlation;2-Medium Correlation;3-Strong Correlation) | | | | | |
|---|---|---|---|---|---|---|
| | CO-1 | CO-2 | CO-3 | CO-4 | CO-5 | CO-6 |
| U5-O1 | 3 | 3 | 1 | - | 3 | 3 |
| U5-O2 | 1 | 1 | - | 2 | 1 | 2 |
| U5-O3 | 2 | 1 | 1 | 1 | 2 | 3 |
| U5-O4 | - | - | 2 | 1 | 2 | 3 |
| U5-O5 | 3 | 3 | 1 | - | 3 | 3 |

## 5.1 FIELD EFFECT TRANSISTOR

A Field Effect Transistor (FET) is a type of transistor that uses an electric field to control the flow of current in a semiconductor channel. The FET is composed of three terminals: the source, the drain, and the gate. The main difference between BJT and FET is that the current through FET is controlled by a voltage applied at the gate terminal. But in BJT, the current through the device is primarily controlled by the current at its base terminal. When a voltage is applied to the gate, an electric field is generated in the insulating layer, which in turn modulates the conductivity of the semiconductor channel. This modulation of conductivity leads to a change in the flow of current between the source and the drain, allowing the FET to act as a switch or an amplifier.

### 5.1.1 Classification of FET

Field Effect Transistors (FETs) can be classified into different categories based on various factors such as their structure, the type of material used, and their operating characteristics. There are two basic types of field effect transistors shown in Fig. 5.1.

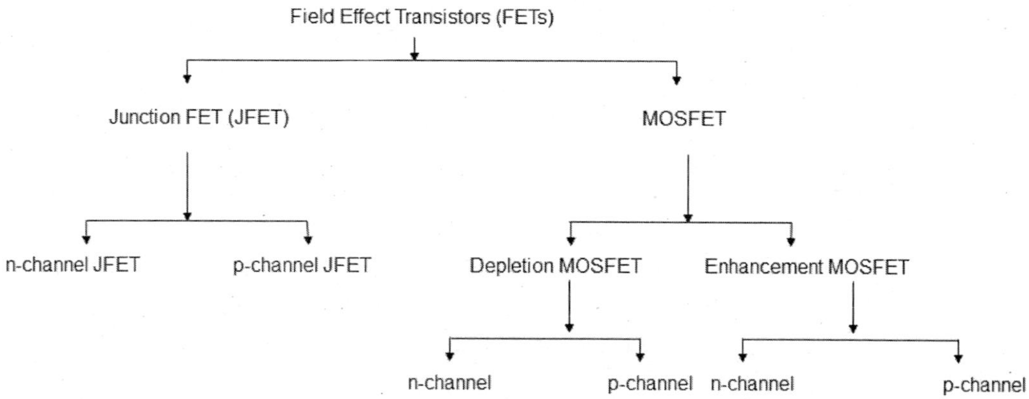


**Fig. 5.1:** Classification of FETs

### 5.1.1.1 Junction Field Effect Transistors (JFETs)

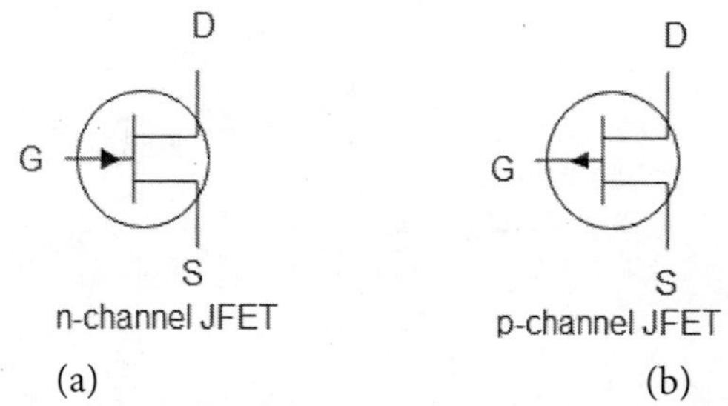


**Fig. 5.2:** Symbol: (a) n-channel, (b) p-channel JFETs

JFETs are the simplest type of FETs and are made by creating a PN junction between the gate and the channel. JFETs are typically used in low-frequency, high-impedance applications. The Fig. 5.2 depicts the symbol for n-channel and p-Channel JFET.

### 5.1.1.2 Metal Oxide Semiconductor Field Effect Transistors (MOSFETs)

MOSFETs are the most widely used type of FET and are made using a metal gate electrode separated from the channel by a thin insulating layer. MOSFETs can be further classified into two types: Enhancement-Mode MOSFETs (E-MOSFETs) and Depletion-Mode MOSFETs (D-MOSFETs). The symbol for n-channel MOSFET under two modes of operations are shown in Fig. 5.3.

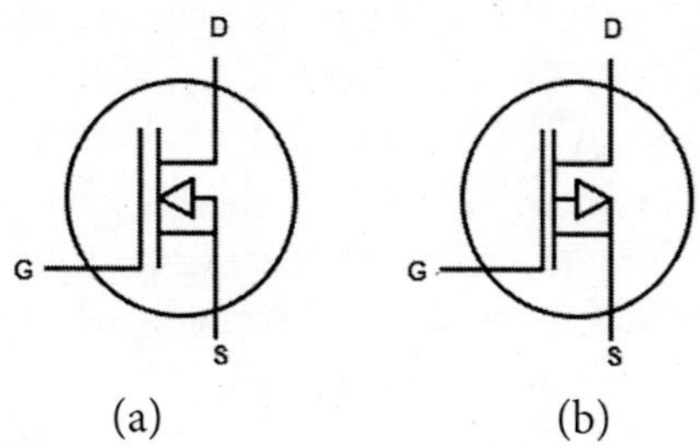


**Fig. 5.3:** Symbol of (a) Depletion mode, (b) Enhancement mode of n-channel MOSFET

### 5.1.2 Comparison of FET with BJT

The Field Effect Transistor (FET) and bipolar junction transistor (BJT) are both types of transistors that can be used as switches or amplifiers in electronic circuits. While both devices have some similarities, they also have some important differences.

**Table 5.1:** Comparison of FET and BJT

| S.No | FET | BJT |
|---|---|---|
| 1. | The FET has a gate terminal, a source terminal, and a drain terminal. | The BJT has a base terminal, an emitter terminal, and a collector terminal. |
| 2. | The FET operates by controlling the width of the conductive channel between the source and drain electrodes using a voltage applied to the gate electrode. | The BJT operates by controlling the current flow between the emitter and collector electrodes. |
| 3. | The FET has low noise characteristics, which makes it suitable for applications that require high sensitivity and low distortion. | The BJT has higher noise characteristics than FETs, which limits the use of applications that require low noise. |
| 4. | The FET has a fast-switching speed. | The BJT has a slower switching speed than FETs. |
| 5. | The FET has a lower temperature sensitivity, which makes it more stable over a wide temperature range. | The BJT has a higher temperature sensitivity, which can affect its performance under varying temperature conditions. |

### 5.1.3 Features of FET

The following are some salient features of FET:

i. **Voltage Controlled Device:** The output current of FET is controlled by the electric field applied at the gate terminal. The current flow between the source and the drain is controlled by a voltage at the third terminal (gate) and so termed as a voltage-controlled current source.

ii. **Unipolar Device:** Only one sort of charge carrier—either electrons or holes is responsible for the current in field effect transistor. Hence it is named as Unipolar Device.

iii. **High Input Impedance:** FETs have a very high input impedance, which makes them ideal for use in high-impedance circuits such as preamplifiers, filters, and sensors.

iv. **Low Noise:** FETs have low noise characteristics, making them useful for low noise applications, such as in audio amplifiers and sensor circuits.

v. **High-Speed Switching:** FETs can switch "on" and "off" very quickly, making them ideal for use in digital circuits such as logic gates and signal processing circuits.

vi. **Low Power Consumption:** FETs require very little power to operate, which makes them ideal for use in battery-operated devices and other low-power applications.

vii. **High Voltage Handling:** FETs can handle high voltages, making them suitable for use in power electronics applications.

viii. **Simple Construction:** FETs have a simple construction, which makes them easy to manufacture and less expensive than some other types of transistors.

ix. **Linear Amplification:** FETs can be used as linear amplifiers, making them useful in a wide range of applications such as audio amplifiers and signal processing circuits.

x. **High Temperature Operation:** Some types of FETs, such as Silicon Carbide (SiC) FETs, can operate at high temperatures, making them suitable for use in high-temperature applications such as power electronics and motor control.

## 5.2 JUNCTION FIELD EFFECT TRANSISTOR (JFET)

The Junction Field Effect Transistor (JFET) is a type of Field Effect Transistor (FET) that is constructed using a combination of doped and undoped semiconductor materials. The JFET has three terminals, the source, gate, and drain. The JFET family includes both n-channel and p-channel. Due to the mobility of its dominant carrier, the n-channel device will be more prominent than p-channel.

### 5.2.1 Construction

Fig. 5.4 illustrates the fundamental design of the n-channel JFET. Remember that the bulk of the structure is composed of the n-type material, which forms the channel between the nested layers of the p-type material. The region doped with p-type semiconductor are connected together to the gate terminal of the JFET. The source terminal is where electrons enter the channel, and the drain terminal is where those electrons exit. The drain and the source are tied to the ends of the n-type channel. The JFET has two p-n junctions when the applied potentials are zero or when there is no bias. Each junction is responsible for developing a depletion zone that mimics the region of a diode under no-bias conditions, as shown in Fig. 5.4. Remember that a depletion region lacks free carriers and cannot support conduction.

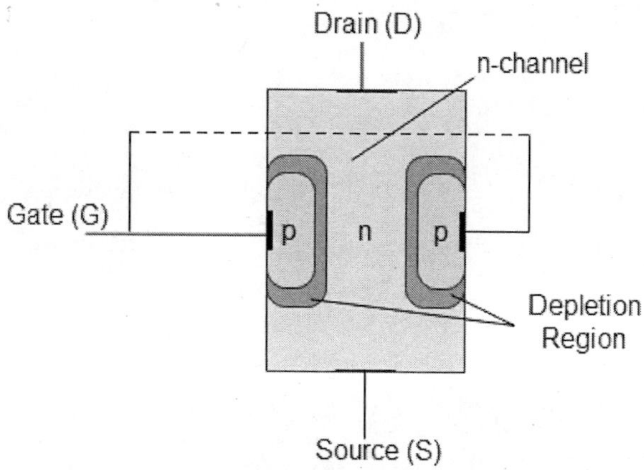


**Fig. 5.4:** Junction Field Effect Transistor (JFET)

The water comparison shown in Fig. 5.5 is a useful way to understand how the JFET control at the gate terminal functions. The water flow through the spigot is analogous to the flow of current through the JFET device, with the water pressure representing the voltage applied across the device from source to drain.

The knob on the spigot can be compared to the gate terminal of the JFET, as turning the knob controls the flow of water, just as changing the voltage at the gate terminal controls the flow of current through the JFET. By reducing the size of the opening in the spigot (i.e. decreasing the gate voltage), the flow of water (i.e. current through the JFET) can be reduced, and vice versa. This comparison can help to visualize the operation of the JFET and understand how changing the gate voltage affects the current flow through the device.

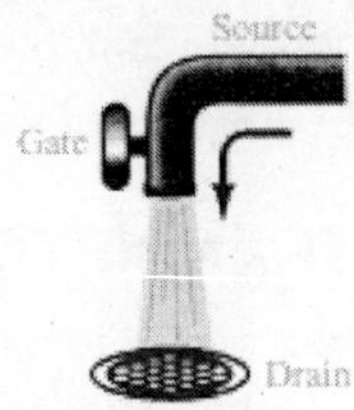


**Fig. 5.5:** Water Analogy for the JFET

## 5.2.2 Working Principle of n-channel JFET

1. **$V_{GS}$ = 0 V, $V_{DS}$ at Some Positive Value**

Consider, a positive voltage $V_{DS}$ is applied across a channel, where $V_{GS}$ = 0 V, and the gate being connected directly to the source. The gate and the source terminals of each p-material are maintained at the same potential. Depletion area is identical to the allocation of the no-bias circumstances as in Fig. 5.4. As soon as the voltage $V_{DD}$ (=$V_{DS}$) is supplied, the conventional current $I_D$ is established with the defined direction of Fig. 5.6. The charge flow path makes it clear that the current flowing via the drain and the source are equal ($I_D = I_S$).

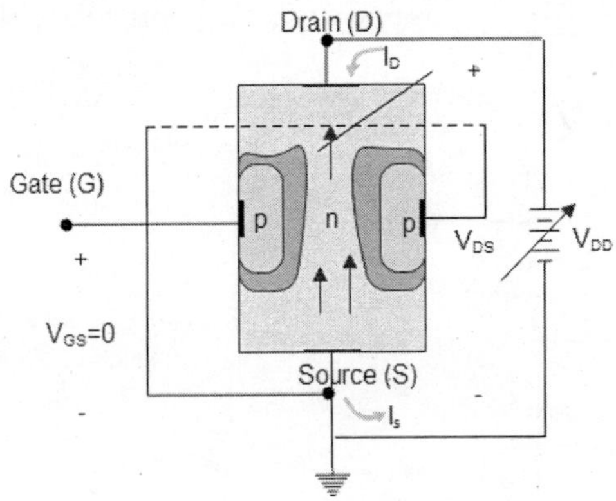


Scan to know more

**Fig. 5.6:** JFET at $V_{GS}$ = 0 and $V_{DS}$ > 0V

The charge flow is mostly unrestricted in this scenario as shown in Fig. 5.6.

2. **When $V_{GS}$ is applied**

The width of the depletion layer is increased. This reduces the width of the conducting channel, there by increases the resistance of n-type bar. The resistance between the drain and source restricts the charge flow in the n-channel JFET.

The channel resistance divides into the sections and the voltage levels over the channel will be established by the current $I_D$, as shown in Fig. 5.7. As a result, the upper portion of the p-type material will experience a reverse bias of about 1.5 V, whereas the lower portion will only experience a reverse bias of 0.5 V. The reverse biasing of the p-n junction throughout the channel, causes a gate current of zero amperes.

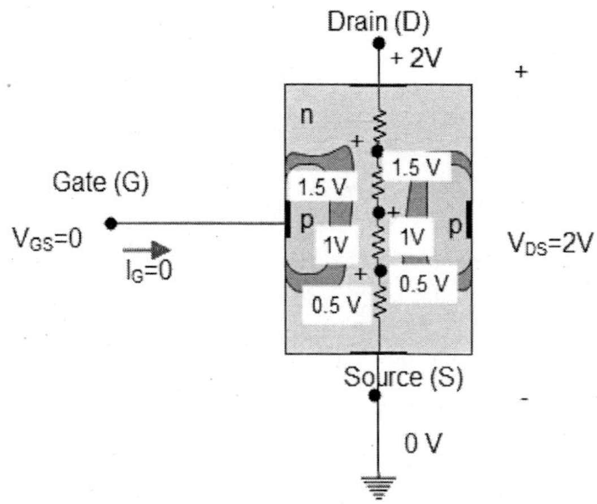


**Fig. 5.7:** Varying reverse-bias potential across the p-n junction of an n-channel JFET

*Note: One significant feature of the JFET is that $I_G = 0$ A.*

## 5.2.3 Characteristics of n-channel JFET

Studying the characteristics of JFET current and voltage are crucial to comprehend the electrical behavior of the device.

The two important characteristics of JFET are drain and transfer characteristics.

### *5.2.3.1 Drain Characteristics*

Fig. 5.8 displays a drain characteristics of n-channel JFET. For various values of $V_{GS}$, the curves show the relationship between the drain current $I_D$ and the drain to source voltage $V_{DS}$.

- **$V_{GS}$ and $V_{DS}$ =0V**

  The channel is completely open at $V_{GS}$=0V. However, no drain current flows through the channel as $V_{DS}$=0V and there is no attractive force for the majority of carriers (electron in n-channel JFET).

- **Pinch-off at $V_{GS}$=0V**

  At $V_{GS}$ =0, the n-type bar behaves as a straightforward semiconductor resistor in response to a low applied voltage $V_{DS}$. The current $I_D$ rises linearly and also the voltage drop throughout the channel rises as $V_{DS}$ is increased. This rise in voltage drop causes the depletion zones to encroach into the channel, narrowing it, and increasing the reverse bias at the gate-source junction. As the drain current $I_D$ increases, the decrease in channel width becomes greater in magnitude. As a result, the rise in $I_D$ in relation to $V_{DS,}$ decreases.

  The channel width is reduced at a certain value of $V_{DS}$, preventing further increase in drain current $I_D$. This is known as Pinch-off voltage ($V_p$). $V_p$ is the drain source voltage at which the current $I_D$ achieves its constant saturation level.

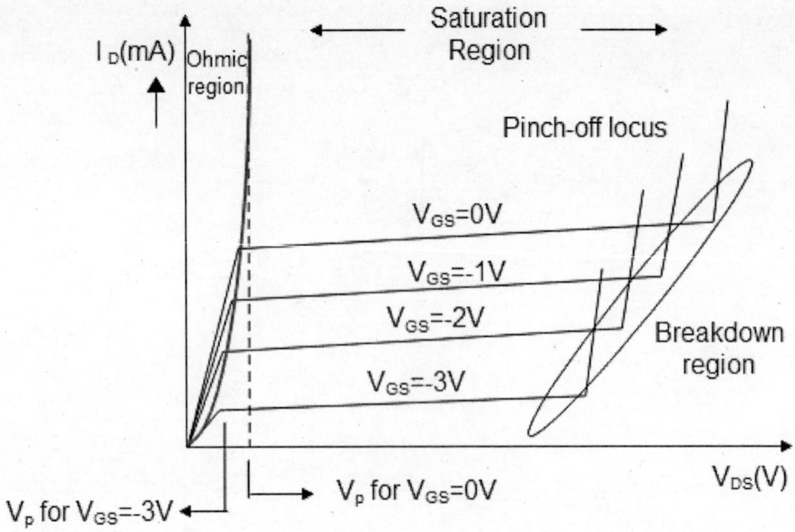


**Fig. 5.8:** Drain V-I Characteristics of n-channel JFET

- **$V_{GS}$ with Negative Bias**
  The gate channel junction is further reverse biased when an external bias of -1V is applied between the gate and the source. It lowers the effective width of the channel that is open to conduction. Therefore, drain current will decrease and pinch off voltage will be attained at a lower value of $I_D$ when $V_{DS}$=0V. A family of curves is produced by applying various negative external bias voltage ($V_{GS}$) values, and it can be seen that for greater negative values of $V_{GS}$, the pinch-off voltage is attained at lower values of $I_D$.
- **Breakdown Region**
  To certain value of $V_{DS}$, the drain current stays constant when the value of $V_{GS}$ increases beyond the pinch-off voltage ($V_p$). Further exceeding $V_{DS}$ will result in the voltage at which the gate-channel connection fails, due to the avalanche effect. At this stage, the device may be damaged as the drain current increases quickly. As the negative gate bias is increased, the value of $V_{DS}$ for breakdown decrease. This is due to the fact that the total reverse breakdown voltage is the sum of the reverse voltage caused by self-pinch-off and the externally applied voltage $V_{GS}$.
- **Ohmic and Saturation Region**
  The JFET's drain characteristics are split into two categories: the saturation region and the ohmic region. The JFET is considered to operate as a voltage-variable resistance when it is in the ohmic zone because the drain current $I_D$ fluctuates with $V_{DS}$. The drain current $I_D$ remains fairly steady and does not change with $V_{DS}$ in the saturation area.
- **Cut-off Region**
  This is also known as the pinch-off region. where the gate voltage, $V_{GS}$ is sufficient to cause the JFET to act as an open circuit as the channel resistance is at maximum.
- **Relation of $V_{GS\,(off)}$ and $V_p$**
  $I_D$ is 0A when $V_{GS}$=-$V_p$

### 5.2.3.2 Transfer Characteristics

Fig. 5.9. illustrates the non-linear relationship between the gate to source voltage $V_{GS}$ and the drain current $I_D$. Shockley's equation provides the relationship between $V_{GS}$ and $I_D$.

$$I_D = I_{DSS}\left(1 - \frac{V_{GS}}{V_P}\right)^2 \tag{5.1}$$

Due to the non-linear relationship between $I_D$ and $V_{GS}$ caused by the squared part of the equation, the curve that result expands exponentially with decreasing $V_{GS}$ magnitudes. The equation 5.1 can be written as,

$$V_{GS} = V_p\left(1 - \sqrt{\frac{I_D}{I_{DSS}}}\right) \tag{5.2}$$

Values of $V_{GS}$ governs $I_D$ in the equation, whereas values of $I_{DSS}$ and $V_p$ are constants. Consequently, this curve displays the JFET's operational limitations, which are,

$$I_D = 0 \text{ when VGS} = V_{GS} = V_{GS(off)} \tag{5.3}$$

$$I_D = I_{DSS} \text{ when } V_{GS} = 0 \tag{5.4}$$

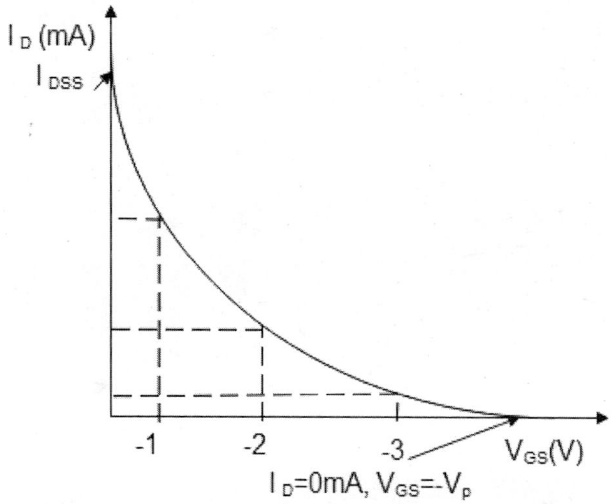


**Fig. 5.9:** Transfer Characteristics of n-channel JFET

## 5.3.4 Construction and Working of p-Channel JFET

The p-channel JFET is built identically like the n-channel JFET, with the p and n-type materials reversed, as shown in Fig. 5.10. The polarities of all current directions and voltages are inverted. Channel width is greatest when $V_{GS} = 0$. The channel width is decreased by raising the positive gate to source ($V_{GS}$) voltage.

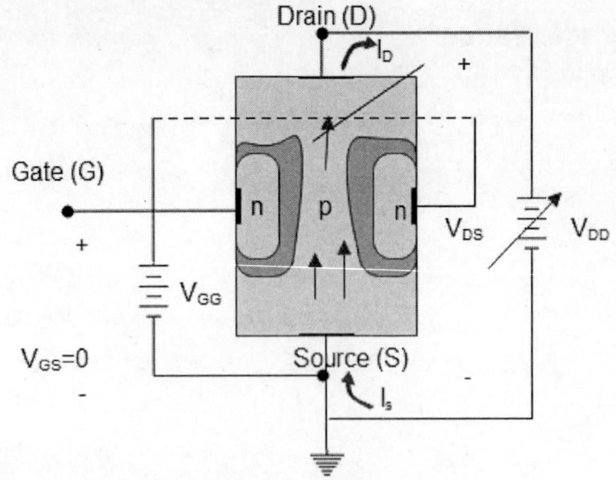


**Fig. 5.10:** p-channel JFET

## 5.2.5 Characteristics of P-Channel JFET

### 5.2.5.1 Drain Characteristics

The source is positive in relation to the drain in a p-channel JFET. The source is the origin of the holes that enter the channel and flow to the drain. By making the source to gate voltage ($V_{SG}$) as negative, the p-n junction diode is reverse biased and the pinch-off is achieved. In Fig. 5.11, the p-channel JFET's drain characteristics is shown.

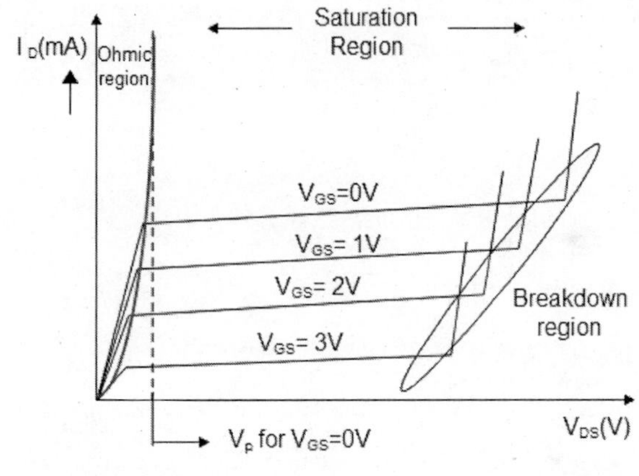


**Fig. 5.11:** Drain I-V Characteristics of p-channel JFET

### 5.2.5.2 Transfer Characteristics

The transfer characteristics of a p-channel JFET is shown in Fig. 5.12. With the exception of the polarities of the $V_{GS}$ and $I_D$, it has characteristics that are similar to those of an n-channel JFET.

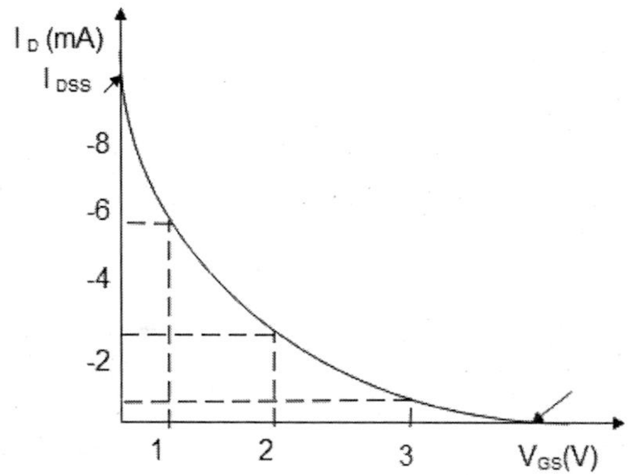


**Fig. 5.12:** Transfer Characteristics of p-channel JFET

**Example 5.1:** Find the value of $V_p$, if the value of drain current is 1.33mA, $I_{DSS}$ = 3mA and the gate-source voltage $V_{GS}$ = –2V..

**Given:** Drain current $I_p$ = 1.33mA, $V_{GS}$ = – 2V and $I_{DSS}$ = 3mA.

**Solution:**

$$I_D = I_{DSS}\left(1 - \frac{V_{GS}}{V_P}\right)^2$$

$$1.33\text{mA} = 3\text{mA}\left(1 - \frac{-2}{V_P}\right)^2$$

$$V_P = -5.970\ \text{V}$$

## 5.2.6 Parameters of JFET

The crucial parameters of JFET are discussed as follows,

### *Transconductance ($g_m$)*

The transconductance $g_m$, is the variation in the drain current for a given variation in the gate to source voltage. Transconductance of JFET is analysed by the use of transfer characteristics shown in Fig. 5.12. Near the top of the curve, $g_m$ has a higher value than at the bottom.

$$\text{Im} = \frac{\Delta I_D}{\Delta BV_{GS}}\bigg|_{V_{DS=\text{constant}}} \tag{5.5}$$

The term "mutual conductance" also refers to $g_m$. With respect to the provided $g_m$, the following equation may be used to determine an approximation of the value for $g_m$ at any location along the transfer characteristic curve.

$$I_m = I_{mo}\left(1 - \frac{V_{GS}}{V_{GS(off)}}\right) \tag{5.6}$$

where $g_{mo}$ is the value of $g_m$ for $V_{GS} = 0$, and it is given by,

$$g_{mo} = -\frac{21_{DSS}}{V_p} \tag{5.7}$$

We know that, $g_{mo} = g_{DSS}\left(1 - \frac{V_{GS}}{V_p}\right)^2$ (5.8)

Differentiating the equation 5.8 with respect to $V_{GS,}$ we get,

$$g_m = \frac{\Delta I_D}{\Delta V_{GS}} = -\frac{21_{DSS}}{V_p}\left(1 - \frac{V_{GS}}{V_p}\right)$$

$$g_m = g_{mo}\left(1 - \frac{V_{GS}}{V_p}\right) \tag{5.9}$$

## Input Resistance and Capacitance

JFET operate with their gate source junction reverse biased. As a result, the input resistance of the gate is extremely high. The JFET has the benefit of having a high input resistance ($R_{IN}$).

$$R_{IN} = \left|\frac{V_{GS}}{I_{GSS}}\right| \tag{5.10}$$

## Drain to Source Resistance

The drain-source resistance $r_d$, is determined from the drain characteristics. When the JFET is functioning in the saturation area. The absolute value of the slope of the drain characteristics curve in the saturation region gives the drain-source resistance ($r_d$). The slope represents the change in drain-source voltage per unit change in drain current.

$$r_d = \frac{\Delta V_{DS}}{\Delta I_D}\bigg|_{V_{GS=constant}} \tag{5.11}$$

## Amplification Factor

The amplification factor, $\mu$ is defined as,

$$\mu = \frac{\Delta V_{DS}}{\Delta V_{GS}}\bigg|_{I_{\eta=constant}} \tag{5.12}$$

$$\mu = \frac{\Delta V_{DS}}{\Delta V_{GS}} = \frac{\Delta V_{DS}}{\Delta I_D} \times \frac{\Delta I_D}{\Delta V_{GS}} r_d \times g_m \tag{5.13}$$

## Power Dissipation

The power dissipation is given by,

$$P_D = I_D \times V_{DS} \tag{5.14}$$

**Example 5.2:** For the reverse voltage of 15V applied to a JFET, the gate current is 10–3μA. Calculate the resistance between the gate and source.

**Given:** Gate to Source Voltage $V_{GS}$ = 15V and gate current IG = $10^{-3}$μA.

**Solution:**

$$\text{Gate to source resistance} = \frac{V_{GS}}{I_G} = \frac{15V}{10^{-6} \times 10^{-8}} = 15 \times 10^9 \Omega = 15000 M\Omega$$

**Example 5.3:** Consider the following information for JFET; $I_{DSS}$ = 2mA, $V_{GS}$ (off) = – 6V and g_m (max) = 5000μS. Find the transconductance for $V_{GS}$ = –4V and calculate $I_D$ at this point.

**Given:** $I_{DSS}$ = 2mA, $V_{GS}$ (off) = –6V and $g_m$ (max) = 5000μS.

**Solution:**

At $V_{GS}$ = 0, the value of $g_m$ is maximum i.e $g_{mo}$

$g_{mo}$ = 5000μS

$$\text{Now, } g_m = g_{mo}\left(1 - \frac{V_{GS}}{V_{GS(off)}}\right)$$

$$= 500\,\mu s\left(1 - \frac{-4}{-6}\right) = 1665\,\mu S$$

$$\text{Also, } I_D = I_{DSS}\left(1 - \frac{V_{GS}}{V_{GS(off)}}\right)^2$$

$$= 2mA\left(1 - \frac{-4}{-6}\right)^2 = 222\mu A$$

**Example 5.4:** Consider the JFET has a drain current of 5mA. If $I_{DSS}$ = 8mA and $V_{GS}$ (off)= – 6V. Calculate i) $V_{GS}$ ii) $V_P$

**Given:** $I_D$ = 5mA, $I_{DSS}$ = 8mA and $V_{GS}$ (off) = –6V

**Solution:**

i. $$I_D = I_{DSS}\left[1 - \frac{V_{GS}}{V_{GS(off)}}\right]^2$$

$$5 = 8\left[1 - \frac{V_{GS}}{6}\right]$$

$$1 + \frac{V_{GS}}{6} = \sqrt{5/8} = 0.790$$

$$V_{GS} = -1.26V$$

ii. $V_P = -V_{GS}$ (off) = 6V.

**Example 5.5:** When the JFET has the value of $V_{GS}$ from -4.1 to -4V, the value of drain current changes from 2 to 2.3 mA. Find the transconductance value.

**Given:** $V_{GS} = -4.1$ to $-4V$, $I_D = 2$ to 2.3mA

**Solution:**

$$\Delta V_{GS} = -4 - (-4.1) = 0.1V$$

$$\Delta I_D = 2.3 - 2 = 0.3mA$$

$$\text{Transconductance, } g_m = \frac{\Delta I_D}{\Delta V_{GS}} = \frac{0.3mA}{0.1V} = \frac{3mA}{IV} = 3000\mu \text{ mho.}$$

## 5.2.7 Small Signal Modeling of JFET

The small signal model of a JFET (Junction Field effect Transistor) is a simplified circuit model that describes the behaviour of the transistor at small signals, typically at frequencies in the audio range or lower.

In the small signal model, the JFET is represented as a voltage-controlled current source, with the gate-source voltage controlling the amount of current flowing through the channel. Using this model, we can analyze the small-signal behavior of the JFET circuit, such as its voltage gain and input impedance, which are important parameters for designing amplifiers and other electronic circuits using JFETs.

The circuit diagram for the small signal model of a JFET is shown Fig. 5.13,

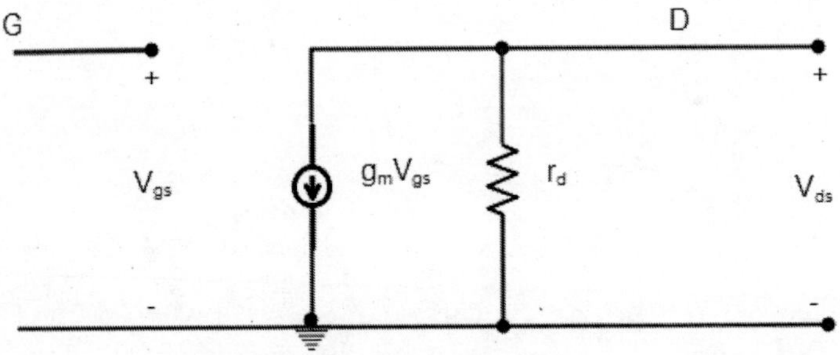


**Fig. 5.13:** Small signal model of JFET

We know that, the drain current $i_D$ is a function of gate source voltage $V_{GS}$ and drain source voltage $V_{DS}$. So, the drain current is written as,

$$i_D = f(V_{GS}, V_{DS}) \tag{5.15}$$

Both the drain and gate currents are varied, the change in drain current is given by,

$$\Delta i_D = \frac{\partial i_D}{\partial V_{GS}}|V_{DS} \ \Delta V_{GS} + \frac{\partial i_D}{\partial V_{DS}}|V_{DS} \ \Delta V_{DS} \tag{5.16}$$

In the small signal notation, the above equation can be written as,

$$i_d = g_m V_{gs} + \frac{1}{r_d} V_{ds} \tag{5.17}$$

where, $g_m = \left.\frac{\partial i_d}{\partial V_{gs}}\right|_{V_{DS}}$ and $r_d = \left.\frac{\partial V_{ds}}{\partial i_d}\right|_{V_{gS}}$

The parameter $g_m$ is known as transconductance defined as the ratio of the change in the small-signal drain current to the change in the small-signal gate-source voltage keeping drain-source voltage constant. Similarly, the parameter $r_d$ is known as drain resistance defined as the ratio of change in drain source voltage to the change in drain current for constant gate source voltage. The amplification factor $\mu$ is defined as the ratio of the change in drain source voltage to the change in gate source voltage.

$$\mu = \frac{\partial V_{ds}}{\partial V_{gs}} = \frac{\partial V_{ds}}{\partial i_d} \cdot \frac{\partial i_d}{\partial V_{gs}} = r_d g_m \quad (5.18)$$

The circuit shown in the Fig. 5.13 satisfies the equation 5.17. Since the input impendence of FET is very high, it is represented by the open circuit at the input terminal. The current $i_d$ is controlled by the gate source voltage. Hence, a current source $g_m V_{gs}$ is connected from drain to source. The current source has its arrow pointing from drain to source.

### 5.2.8 Comparison of n-channel JFET and p-channel JFET

The comparison of JFET and MOSFET depends on the symbol and the characteristics are summarized in Table 5.2.

**Table 5.2:** Comparison between p-channel JFET and n-channel JFET

| S.No | p-channel JFET | n-channel JFET |
|---|---|---|
| 1. | In p-channel JFET, the current carriers are holes. | In n-channel JFET, the current carriers are electrons. |
| 2. | The channel is made of p-type material. | The channel is made of n-type material. |
| 3. | The symbol of p-channel shows the arrow pointing away from source/drain channel. | The symbol of n-channel shows the arrow pointing towards the drain/source channels. |
| 4. | Noise is very high compared to n-channel JFET. | Noise is very less than that of p-channel JFET. |
| 5. | Transconductance is very less compared to n-channel JFET. | Transconductance is larger in n channel JFET. |

## 5.3 MOS CAPACITOR

A MOS (Metal-Oxide-Semiconductor) capacitor is a type of electronic device that consists of a metal electrode, an insulating oxide layer, and a semiconductor region. The oxide layer serves as a dielectric that separates the metal electrode from the semiconductor region, and it is typically made of materials such as Silicon dioxide ($SiO_2$) or Aluminium oxide ($Al_2O_3$). The semiconductor layer can be n-type or p-type and acts as one plate of the capacitor, while the metal gate electrode serves as the other plate.

The MOS capacitor seen in Fig. 5.14 serves as the brain of the MOSFET. The thickness ($t_{ox}$) and permittivity ($\varepsilon_{ox}$) of the oxide layer are important parameters in determining the capacitance and performance of the MOS capacitor.

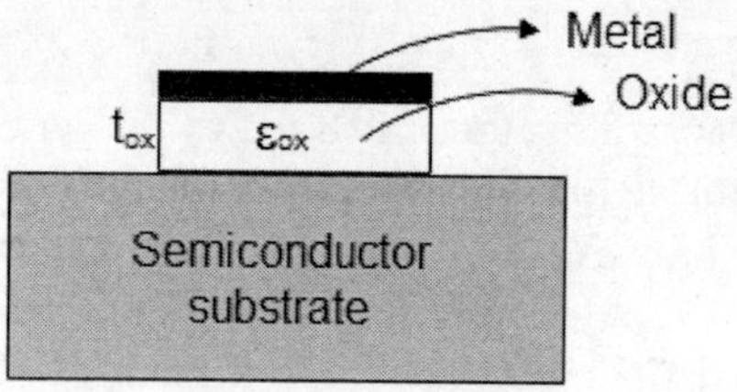


**Fig. 5.14:** MOS Capacitor

### 5.3.1 Energy-Band Diagram

The straightforward parallel-plate capacitor can be used to more clearly illustrate the mechanics of the MOS construction. As demonstrated in Fig. 5.15, a parallel-plate capacitor has a top plate that is negatively charged compared to the bottom plate. The two plates are separated by an insulating substance. As a result of this bias, an electric field is formed between the two plates. The capacitance per unit area for this geometry is defined as,

$$C = \frac{\varepsilon}{d} \tag{5.19}$$

where ε is the permittivity of the insulator, the permittivity for $SiO_2$ is 3.9 and d is the distance between two parallel plates. The charge stored in the capacitor is expressed as,

$$Q = C.V \tag{5.20}$$

The field applied on the plate is given by,

$$E = \frac{V}{d} \tag{5.21}$$

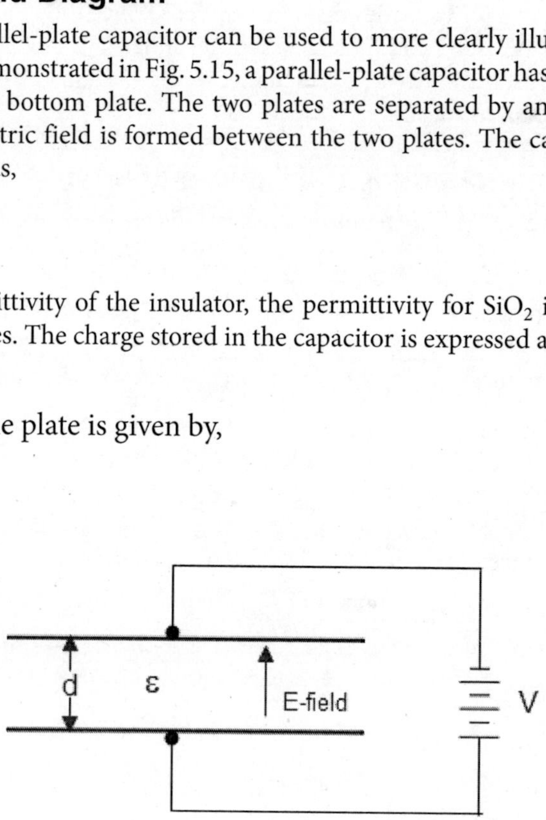


**Fig. 5.15:** Parallel Plate Capacitor

The energy band diagram of a MOS (Metal-Oxide-Semiconductor) capacitor shows the distribution of energy levels in the device under equilibrium conditions, i.e., without any applied bias voltage. The diagram illustrates the energy levels of the semiconductor and metal regions, separated by the oxide layer. The energy-band diagrams of the p-type substrate MOS capacitor is shown in Fig. 5.16. Here, $E_c$ is the conduction band edge, $E_V$ is the valance band edge, $E_F$ is the fermi level and $E_{Fi}$ is the intrinsic fermi level. The Fig. 5.16 shows the ideal situation where no bias is applied to the MOS device. The energy bands are flat in the semiconductor region, indicating that there is no net charge present in the device under equilibrium conditions.

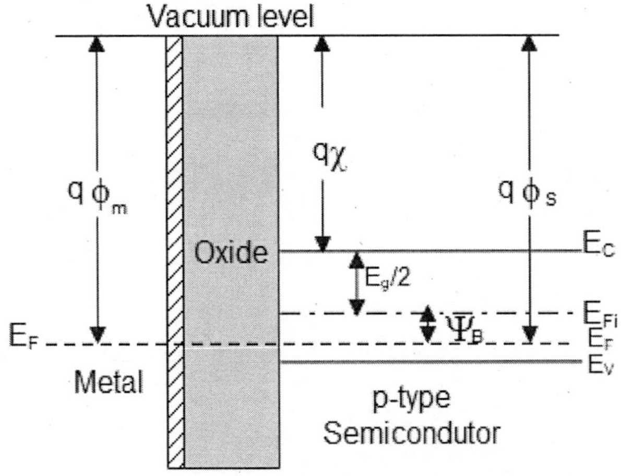


**Fig. 5.16:** Energy Band Diagram of a MOS Capacitor with a p-type substrate

where, X is electron affinity, $\Phi_m$ is metal work function and $\Phi_s$ is semiconductor work function, The energy gap between $E_c$ and $E_i$ is denoted by $E_g/2$ and the gap between $E_F$ and $E_i$ is denoted by $\Psi_B$.

## 5.3.2 Ideal C-V Characteristics

The energy-band diagram of the MOS capacitor with a p-type substrate for various gate biases are discussed below. The MOS capacitor has three operational circumstances.

i. Accumulation Region
ii. Depletion Region
iii. Inversion Region

**i. Accumulation Region**

In the *accumulation region* of operation, a negative voltage is applied to the gate electrode, if the substrate is p-type. In the energy-band diagram in Fig. 5.17 (a), the valence-band edge is closer to the Fermi level at the heterojunction because there are more holes than in the bulk material.

$$C_{acc} = C_{ox} = \frac{\varepsilon_{ox}}{t_{ox}} \tag{5.22}$$

The accumulation of holes at the interface results in a capacitor dominated by oxide capacitance. In this mode of operation, only the oxide capacitance is present since there is no depletion region in the semiconductor.

**ii. Depletion Region**

When a positive voltage is applied to the gate electrode, the energy-band diagram of the MOS system is shown in Fig. 5.17 (b). As a result of the positive voltage on the gate, electrons are repelled from the interface, resulting in a depletion region between the oxide and semiconductor. This depletion region creates a potential barrier that repels the majority carriers (electrons in this case) away from the interface. The depletion region extends into the semiconductor, resulting in a space-charge region with a width of $x_d$. As shown in the energy-band diagram, the conduction- and valence-band edges bend, creating a potential gradient across the space-charge region. The intrinsic Fermi level (Ei) and conduction band edge approach the Fermi level due to the presence of fixed charges in the oxide and the electric field in the depletion region. The potential $\Psi_B$ is the difference between $E_{Fi}$ and $E_F$ is given by,

$$\psi_B = V_T \ln (N_a/n_i) \tag{5.23}$$

where $N_a$ is the acceptor doping concentration and $n_i$ is the intrinsic carrier concentration.

The depletion layer width or space charge region is given by

$$x_d = \left(\frac{2\varepsilon_s \psi_s}{eN_a}\right)^{1/2} \tag{5.24}$$

The potential $\psi_s$ is the surface potential, it is the difference (in V) between EFi measured in the bulk semiconductor and $E_{Fi}$ measured at the surface.

Depletion capacitance is formed due to the formation of a depletion region under the small positive voltage.

$$1/C_{dep} = 1/C_{ox} + 1/C_{SD} \tag{5.25}$$

$$C_{dep} = \frac{C_{ox}C_{SD}}{C_{ox} + C_{SD}} \tag{5.26}$$

This condition gives the minimum capacitance value.

iii. **Inversion Region**

In the *inversion mode* of operation, a large positive voltage is applied to the gate of the MOS capacitor, causing the surface of the semiconductor material to become enriched with free electrons and transforming it into an n-type region. This results in the formation of an inversion layer at the oxide-semiconductor interface, which acts as a conducting channel between the source and drain regions. Weak inversion occurs when the gate voltage is just sufficient to cause a small number of minority carriers to accumulate in the channel region is shown on Fig. 5.17 (c). While, strong inversion occurs when the gate voltage is high enough to create a large number of majority charge carriers (electrons or holes) in the channel. The density of charge carriers in the channel is high, and the channel resistance is low.

If the inversion charge may react to a change in capacitor voltage, the capacitance is once more merely the oxide capacitance, or it is defined as,

$$C_{inv} = C_{ox} = \frac{\varepsilon_{ox}}{t_{ox}} \tag{5.27}$$

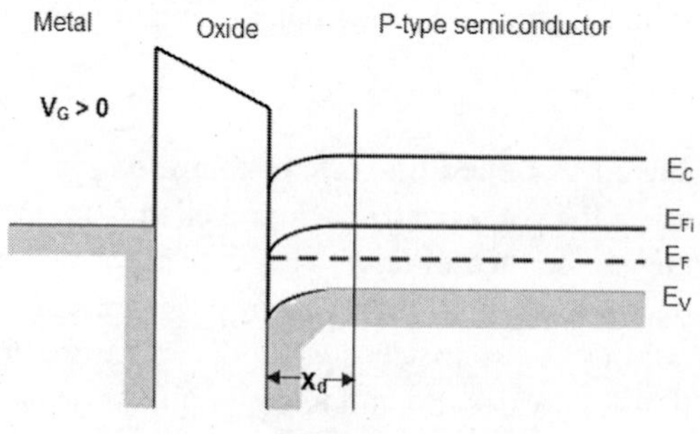


(a)

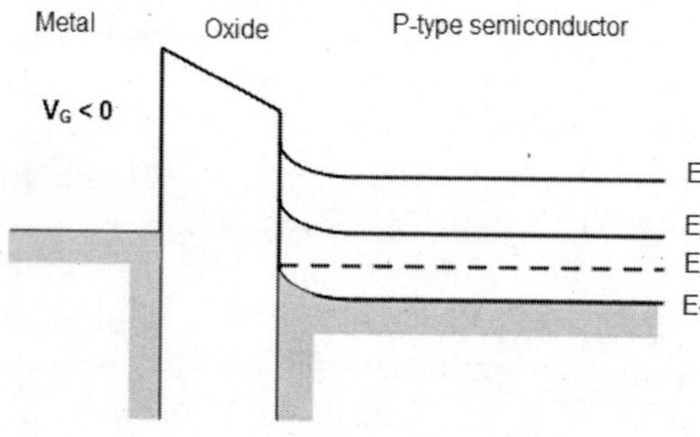


(b)

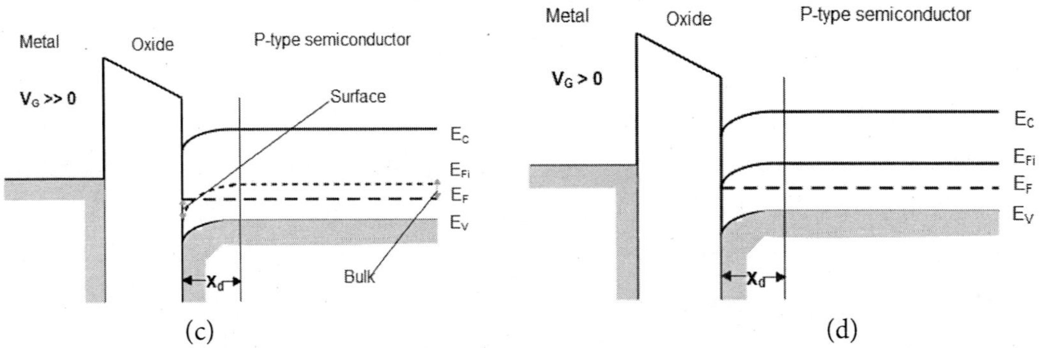


**Fig. 5.17:** Band Diagram of a) Accumulation Region, b) Depletion Region, c) Weak Inversion, d) Strong Inversion

The energy bands for the scenario are depicted in Fig. 5.17 (d). In this scenario, the electron concentration at the surface is identical to the hole concentration of the bulk that is $\psi_S = 2\psi_B$.This point is known as *threshold inversion point*. If the gate voltage climbs over this threshold level, the conduction band will budge somewhat closer to the Fermi level, but this movement in the conduction band at the surface is only a minor function of the gate voltage. However, the surface potential has an exponential relationship with the surface electron concentration. The electron concentration may change by orders of magnitude when the surface potential increases by a few (kT/e) volts. The space charge zone has so virtually achieved its maximum width in this instance. The maximum depletion width will be,

$$X_d = \left(\frac{4\varepsilon_s \psi_B}{eN_a}\right)^{1/2} \tag{5.28}$$

Fig. 5.18 depicts the optimal C-V characteristics of the MOS capacitor with a p-type substrate. The capacitance-voltage characteristics of a MOS capacitor in the accumulation region shows a linear behavior since the capacitance is proportional to the voltage applied to the gate electrode. The slope of the C-V curve is proportional to the oxide capacitance and inversely proportional to the thickness of the oxide layer. As the oxide thickness increases, the slope of the C-V curve decreases, and the capacitance at a given voltage decreases as well. The slope of the C-V curve in the depletion region is inversely proportional to the width of the depletion region, which is proportional to the square root of the voltage applied to the gate. Therefore, the slope of the C-V curve decreases as the voltage applied to the gate increases, resulting in a smaller change in capacitance for a given change in voltage.

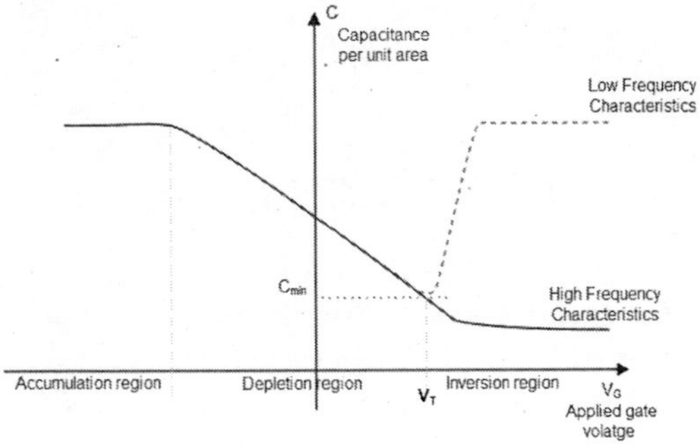


**Fig. 5.18:** C-V Characteristics of MOS Capacitor

**Example 5.6:** The energy band diagram of a MOS Capacitor device is sketched in the figure below. Assume that the electrostatic potential is zero in the semiconductor bulk, (i.e., at large distance from Si-SiO2 interface) and that there is no metal-semiconductor work function difference. Assume the relative dielectric constant of the oxide to be $\varepsilon_{ox} = 3.9$. (Take ni = $10^{10}$ cm$^{-3}$), kT = 26 meV, $E_g = 1.1$ eV, $\varepsilon_{si} = 11.8$).

Calculate the value of $\phi_F$, surface potential $\phi_s$, find the applied gate voltage $V_G$ and the doping density $N_d$.

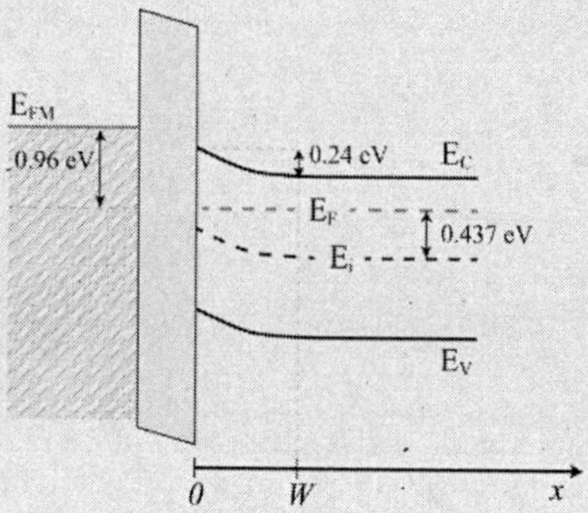


**Given:** $n_i = 10^{10}$ cm$^{-3}$, kT = 26 meV, $E_g = 1.1$ eV, $\varepsilon_{si} = 11.8$, and $\varepsilon_{ox} = 3.9$.

**Solution:** Let K = 1.38 × $10^{-23}$) J/K, T=300K.

$$\phi_F = E_i - E_F = -0.437\text{eV}$$

Surface potential,

$$\phi_s = E_i - E_{Surface} = -0.24\text{eV}$$

Gate voltage applied = –0.96eV- Metal fermi level moves up. Since the difference is 0.96eV the applied voltage must be –0.96V

Doping density is calculated using,

$$\phi_F = \frac{kT}{q} \text{In}\left(\frac{N_d}{ni}\right)$$

$$0.437 = 0.026 \text{ In}\left(\frac{N_d}{10_{10}}\right)$$

$$N_d = 2 \times 10^{17} \text{cm}^{-3}$$

## 5.4 METAL OXIDE SEMICONDUCTOR FIELD EFFECT TRANSISTOR (MOSFET)

Metal Oxide Semiconductor Field Effect Transistor or MOSFET for short serves as an alternative to the Junction Field Effect Transistor (JFET). MOSFET is a voltage-controlled field effect transistor. It is utilized in a wide range of electronic circuits. It contains 4 terminals including, Gate (G), Source (S), Drain (D), and Body / Substrate (B).

In MOSFET, between gate and substrate, a $SiO_2$ oxide layer is present. The gate is made with a metal conductor, the oxide layer is an insulator and the substrate is made with a semiconductor. Thus, from gate terminal to substrate it contains a metal oxide semiconductor, so it is named as MOSFET. Here the gate is insulated from the channel. So MOSFET is also called Insulated Gate Field Effect Transistor (IGFET).

Consider JFET, for n-channel JFET it has only negative gate operations and for p-channel JFET it has only positive gate operations. Due to this, the channel width can be decreased from its zero-bias condition. This is called a depletion mode operation. So JFET can operate only in, gate reverse biased condition and depletion mode. This is the main drawback of JFET. For this drawback, the better replacement is MOSFET. MOSFET can work in two operation modes–depletion mode and enhancement mode. The major advantage of MOSFET is high input impedance and low cost of production over JFET.

## 5.4.1 Types of MOSFET

Based on the construction and operating modes MOSFET can be classified into two types. They are,

- Depletion MOSFET or D-MOSFET
- Enhancement MOSFET or E-MOSFET

In a D-MOSFET, the channel is diffused between the source and drain in its basic structure itself. So, depletion mode and enhancement mode are two possible modes of operation in D-MOSFET. In E-MOSFET, there is no continuous channel, and the channel is induced during E-MOSFET operation. So, only the enhancement mode can be employed when operating an E-MOSFET.

### *Depletion MOSFET (D-MOSFET)*

Depletion MOSFET is further classified as n-channel D-MOSFET and p-channel D-MOSFET. The circuit symbols of n-channel and p-channel D-MOSFET is shown in Fig. 5.19 (a) and Fig. 5.19 (b).

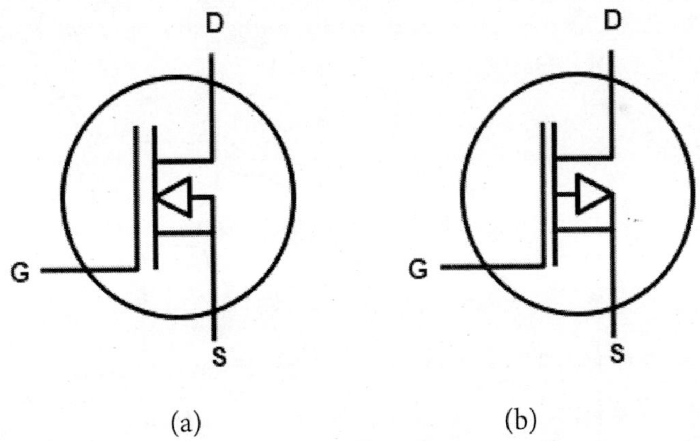


**Fig. 5.19:** Circuit Symbol of (a) n-channel D-MOSFET, (b) p-channel D-MOSFET

The simplified representation of n-channel and p-channel D-MOSFET is shown in Fig. 5.20 (a) and Fig. 5.20 (b). In D-MOSFET, a channel is present between source and drain terminal continuously in its structure. So, in its circuit symbol and simplified representation, near the gate region, there is a continuous line present.

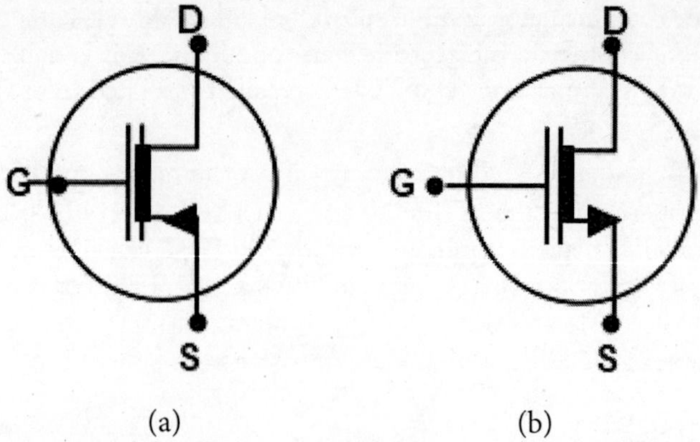


**Fig. 5.20:** Simplified representation of (a) n-channel D-MOSFET, (b) p-channel D-MOSFET

## 5.4.2 Construction of n-channel Depletion MOSFET

The basic construction of n-channel depletion MOSFET is depicted in Fig. 5.21. The source and drain are made with highly doped n-regions, the substrate is made with lightly doped p-regions and the channel is made with n-regions. A thin layer of Silicon dioxide ($SiO_2$) is present in between gate and channel, to provide insulation. So, in MOSFET, there is no direct electrical connection between the gate and the channel. The source terminal is internally connected with substrate. Gate, source and drain terminals are connected via metallic contact to its corresponding doped regions.

The gate and the channel regions form a parallel-plate capacitor and oxide layer act as dielectric. So, changing the gate voltage causes the changes in capacitor electric field and it leads to n-channel resistance changes. In D-MOSFET both positive and negative gate voltage is possible. The negative gate voltage operation is called as depletion mode and the positive gate voltage operation is called as enhancement mode.

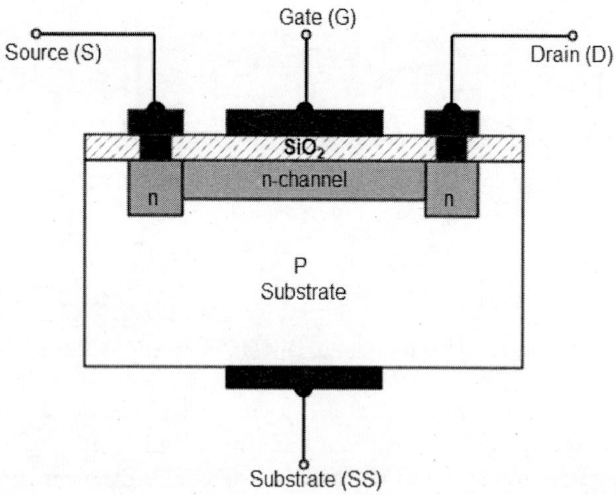


**Fig. 5.21:** n-channel Depletion MOSFET

## 5.4.3 Working Principle of n-channel Depletion MOSFET

### *5.4.3.1 Depletion Mode in n-Channel D-MOSFET*

The operation of n-channel D-MOSFET in depletion mode is shown in Fig. 5.22.

i. Consider gate to source voltage, $V_{GS}$=0V. By applying drain to source voltage ($V_{DS}$), the drain terminal attracts the free electrons in the n-channel. So, there is a current flow in the circuit from source to drain, which is called drain-source leakage current ($I_{DSS}$).

ii. Consider the negative gate voltage applied. The electrons in the n-channel are depleted and the holes in the p-substrate are attracted towards the gate region and recombination takes place. So, the electrons for current conduction in channel are decreased and the channel conductivity also decreases. Finally, the drain current decreases in depletion mode operation.

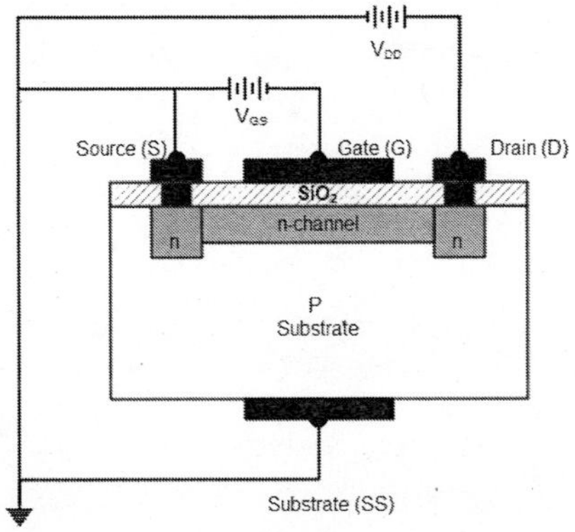


**Fig. 5.22:** Depletion Mode operation of n-channel D-MOSFET

### *5.4.3.2 Enhancement Mode in n-Channel D-MOSFET*

The operation of n-channel D-MOSFET in enhancement mode is shown in Fig. 5.23. Consider the positive gate voltage applied. Due to reverse leakage current, the electrons in the p-substrate are also attracted towards the gate region. So, the channel is enhanced or increased. The electrons for current conduction in the channel are increased and the channel conductivity also increases. Finally, the drain current increases in enhancement mode operation.

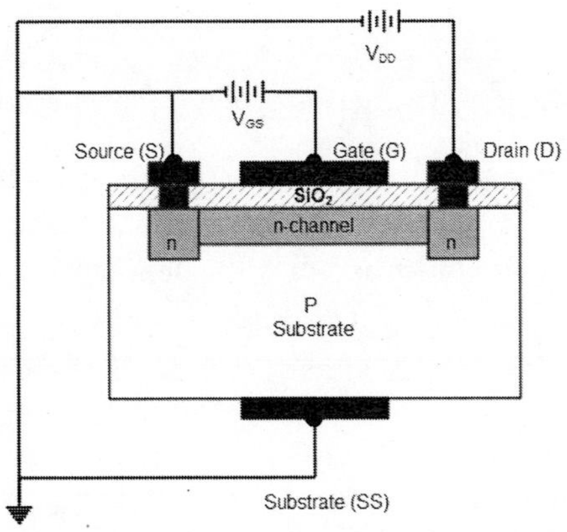


**Fig. 5.23:** Enhancement mode operation of n-channel D-MOSFET

### 5.4.3.3 Transfer and Drain Characteristics

The transfer characteristics of n-channel depletion MOSFET is shown in Fig. 5.24 (a). The characteristics of the device is explained as,

i. If $V_{GS}$=0V, then drain current is equal to drain-source leakage current. So, $I_D=I_{DSS}$.
ii. When $V_{GS}$ is negative, $I_D$ starts decreasing and reaches zero at $V_{GS}=V_{GS\,(off)}$. This is called depletion mode.
iii. When $V_{GS}$ is positive, $I_D$ starts to increase and the maximum value of $I_D$ is obtained from the datasheet of D-MOSFET. This is called enhancement mode.

The transconductance curve of the D-MOSFET is similar to the curve of the JFET. The drain current equation is given as,

$$I_D = I_{DSS}\left(1-\frac{V_{GS}}{V_{GS(off)}}\right)^2 \tag{5.29}$$

The transconductance of the D-MOSFET is given as,

$$g_m = g_{mo}\left(1-\frac{V_{GS}}{V_{GS(off)}}\right) \tag{5.30}$$

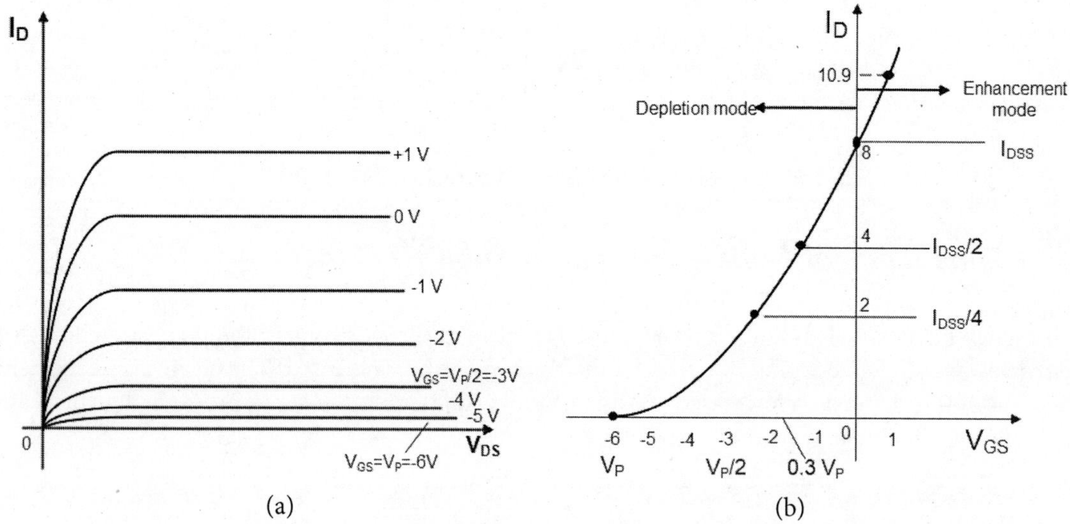


**Fig. 5.24:** (a) Drain, (b) Transfer Characteristics for an n-channel depletion-type MOSFET

The drain characteristics of n-channel depletion MOSFET is shown in Fig. 5.24(b). The drain characteristics have a cut-off region, ohmic region and saturation region.

i. The region below $V_{GS} = V_P$ is known as, cut-off region. Ideally, the drain current in the cut-off region is zero.
ii. For particular $V_{GS}$, drain current ($I_D$) increases with respect to drain to source voltage ($V_{DS}$) is known as the ohmic region.
iii. For particular $V_{GS}$, drain current ($I_D$) constant with respect to drain to source voltage ($V_{DS}$) is known as the saturation region. The ratio of change in drain current with respect to change in gate to source voltage with constant drain to source voltage is known as transconductance ($g_m$) and it is given by,

$$g_m = \left(\frac{\nabla I_D}{\Delta V_{GS}}\right)_{\text{constant } V_{DS}} \qquad (5.31)$$

### 5.4.4 P-channel Depletion MOSFET

The basic construction of p-channel depletion MOSFET is depicted in Fig. 5.25. The construction of p-channel D-MOSFET is exactly opposite of n-channel D-MOSFET. The source and drain are made with highly doped p-regions, the substrate is made with lightly doped n-regions and the channel is made with p-regions. Here, the positive gate voltage operation is called depletion mode and the negative gate voltage operation is called enhancement mode. Remaining all operations are the same as n-channel D-MOSFET.

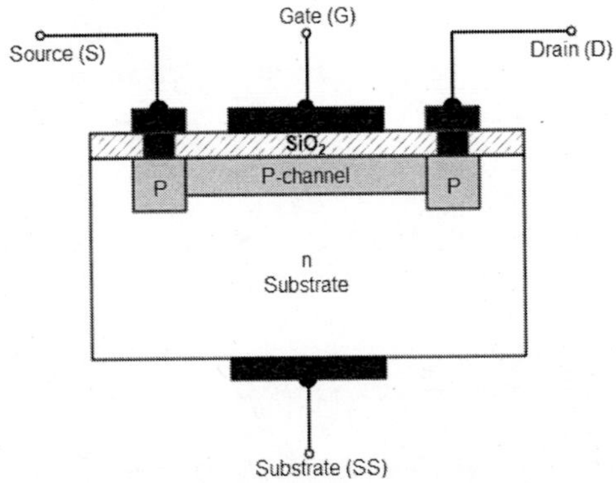


**Fig. 5.25:** P-channel Depletion MOSFET

The transfer characteristics of P-channel depletion MOSFET is shown in Fig. 5.26 (a). It is a mirror image of an n-channel D-MOSFET shown in Fig. 5.24 (a). The drain characteristics of p-channel depletion MOSFET is shown in Fig. 5.26 (b). Compared to n-channel D-MOSFET, p-channel D-MOSFET drain characteristics gate to source voltage ($V_{GS}$) polarity is changed.

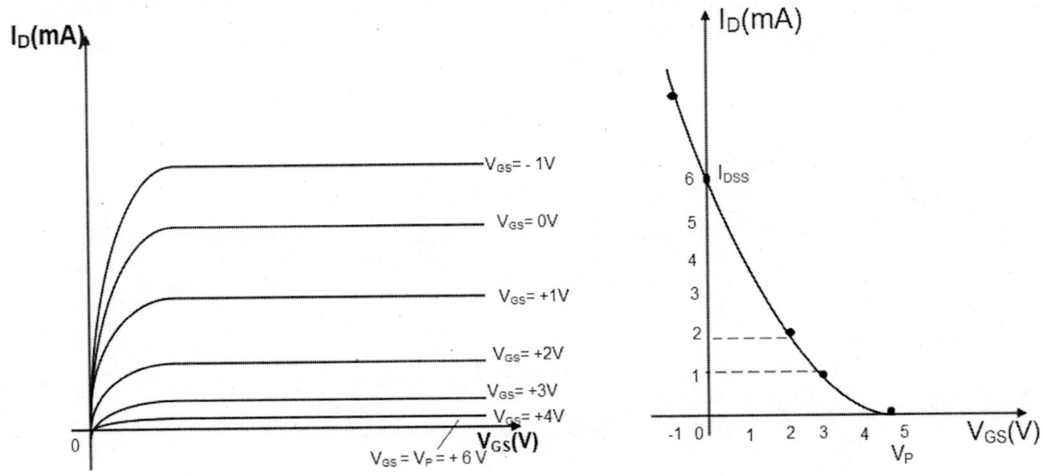


(a) (b)

**Fig. 5.26:** (a) Transfer Characteristics, (b) Drain Characteristics for an p-channel D-MOSFET.

## 5.4.5 Enhancement MOSFET (E-MOSFET)

In enhancement MOSFET or E-MOSFET, it does not have a physical channel between source and drain terminal; E-MOSFET has only enhancement mode, so there is no depletion mode in it.

### 5.4.5.1 Types and Circuit Representation of E-MOSFET

Enhancement MOSFET is further classified as n-channel E-MOSFET and p-channel E-MOSFET. The circuit symbol of n-channel and p-channel E-MOSFET is shown in Fig. 5.27 (a) and Fig. 5.27 (b).

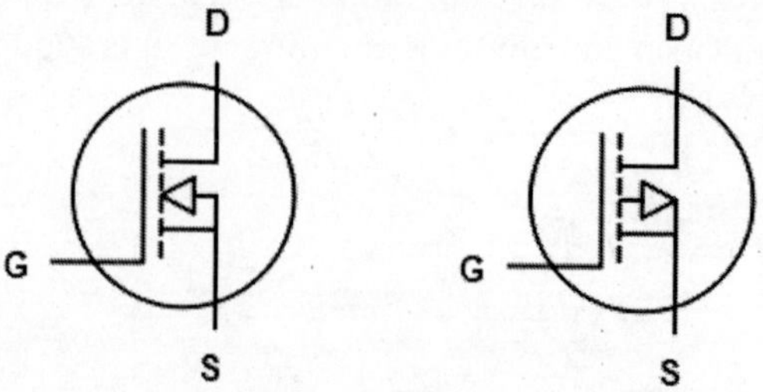


(a) (b)

**Fig. 5.27:** Circuit Symbol of (a) n-channel E-MOSFET, (b) p-channel E-MOSFET

The simplified representation of n-channel and p-channel E-MOSFET is shown in Fig. 5.28 (a) and Fig. 5.28 (b). In E-MOSFET, no continuous channel is present.

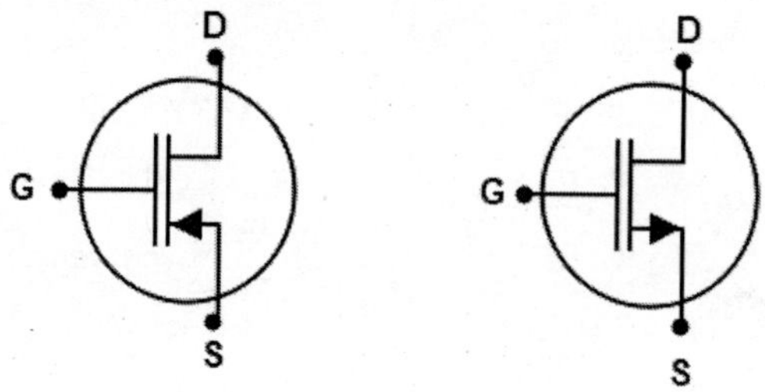


(a) (b)

**Fig. 5.28:** Simplified Representation of (a) n-channel E-MOSFET, (b) p-channel E-MOSFET

### *5.4.5.2 Construction and Working of n-channel E-MOSFET*

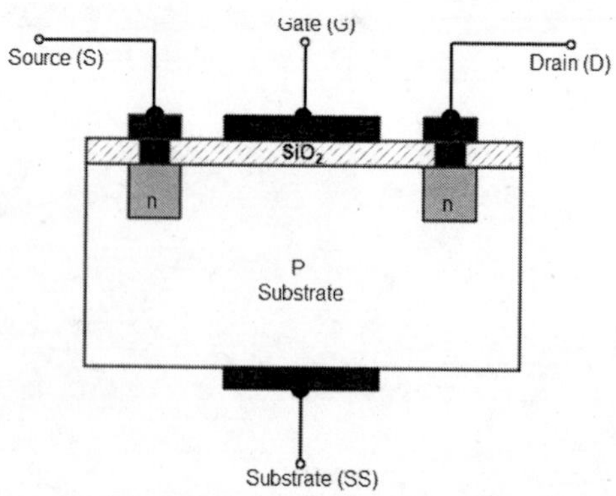


**Fig. 5.29:** n-channel Enhancement MOSFET

The basic construction of n-channel enhancement MOSFET is depicted in Fig. 5.29. Like n-channel depletion MOSFET, the source and drain are made with highly doped n-regions and the substrate is made with lightly doped p-regions. Here channel region is absent, that is no physical channel is present in between source and drain region. A thin layer of silicon dioxide ($SiO_2$) is present in between gate and substrate, to provide insulation. So, in MOSFET, there is no direct electrical connection between the gate and the substrate. The source terminal is internally connected with substrate. Gate, source and drain terminals are connected via metallic contact to its corresponding doped regions. In E-MOSFET, only positive gate voltage is possible. So, it is called an enhancement MOSFET. The operation of n-channel E-MOSFET in depletion mode is shown in Fig. 5.30.

Consider gate to source voltage, $V_{GS}$=0V. There is no connecting channel between source and drain terminal. Only a smaller number of thermally produced free electrons available at p-substrate. So, drain current is ideally zero ($I_D$=0). This operation is completely different from JFET and D-MOSFET.

i. Consider the positive gate to source voltage ($V_{GS}$) applied. The gate region attracts the electrons in the p-substrate. These electrons combine with the free holes present under $SiO_2$. Once the holes are filled, the free electrons are accumulated near the $SiO_2$ layer thus forming an n-channel in E-MOSFET. Free electrons are starts to move from source to drain region, the drain current ($I_D$) flows from drain to source and the n-channel E-MOSFET is turned ON. The minimum required positive $V_{GS}$ to turn ON the E-MOSFET is known as threshold voltage ($V_T$).
ii. Consider the gate to source voltage ($V_{GS}$) is greater than threshold voltage ($V_T$). The channel is enhanced or increased. The electrons for current conduction in the channel are increased and the channel conductivity also increases. Finally, the drain current increases as $V_{GS}$ increases.
iii. Consider the gate to source voltage ($V_{GS}$) is constant and greater than the threshold voltage ($V_T$) and drain to source voltage ($V_{DS}$) is increased. The drain current is increasing with respect to $V_{DS}$ up to Pinch-off voltage ($V_P$). At $V_{DS} = V_P$, the channel region is pinched-off, so the drain current gets saturated ($I_D$ becomes constant).

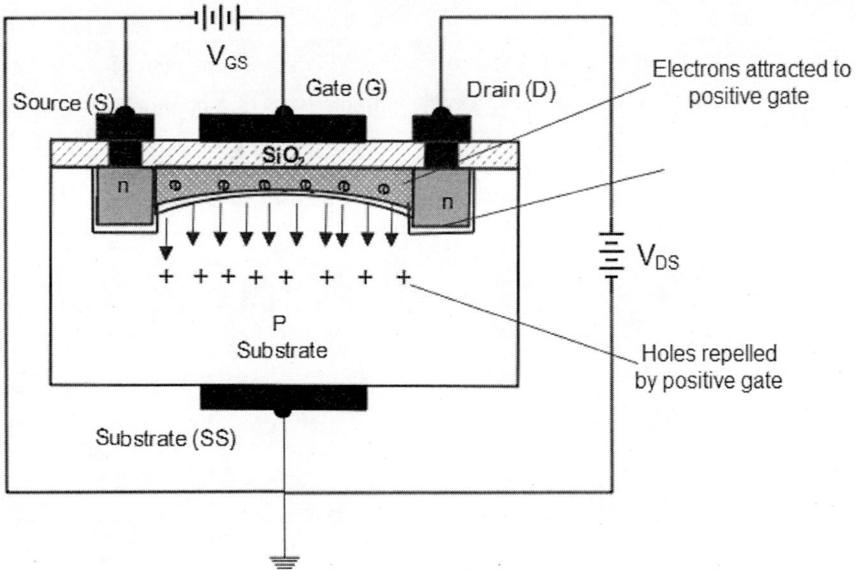


**Fig. 5.30:** Channel Formation in the n-channel E-MOSFET

### 5.4.5.3 Transfer and Drain Characteristics of n-channel E-MOSFET

The transfer characteristics of n-channel enhancement MOSFET is shown in Fig. 5.31(a). The characteristics of the device is explained as,

i. If $V_{GS}$ = 0V, then drain current is ideally zero.

ii. When $V_{GS} > V_T$, $V_{DS}$ = 0V, the channel starts to form, no drain current flows.

iii. When $V_{GS}$ is increased further and $V_{GS} > V_T$, $V_{DS}$ is also increased $I_D$ starts to increase.

The drain current equation of E-MOSFET is given as,

$$ID = K\,(V_{GS} - V_T)^2 \tag{5.32}$$

Here K is the conduction parameter and is given by,

$$K = \frac{1}{2}\mu_n C_{ox}\frac{W}{L} \tag{5.33}$$

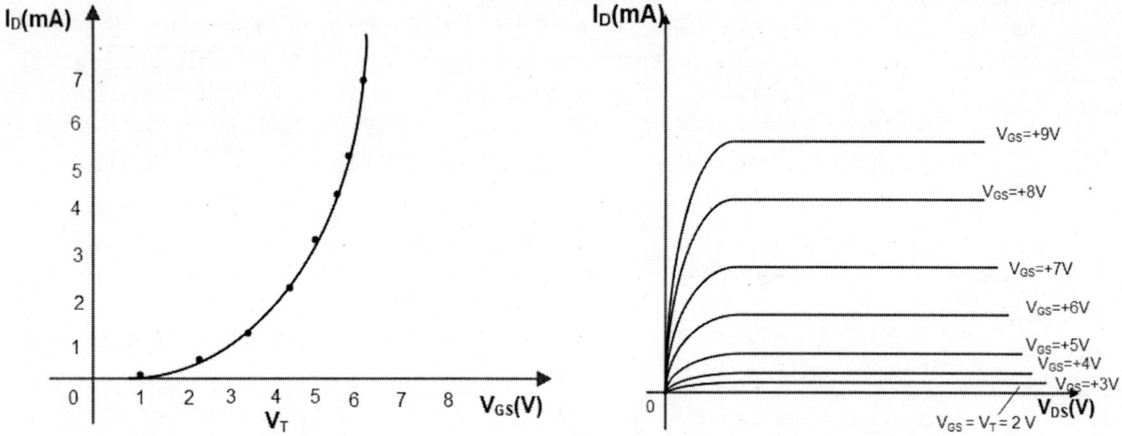


(a) (b)

**Fig. 5.31:** (a) Transfer Characteristics, (b) Drain Characteristics of n-channel E-MOSFET

The drain characteristics of n-channel enhancement MOSFET is shown in Fig. 5.31(b). The drain characteristics have cut-off region, linear region and saturation region.

i. The region below $V_{GS} < V_T$ is known as, cut-off region. Ideally, the drain current in the cut-off region is zero.

ii. For $V_{GS} > V_T$ and $V_{DS} < V_P$, drain current ($I_D$) increases with respect to drain to source voltage ($V_{DS}$) is known as the linear region.

iii. For $V_{GS} > V_T$ and $V_{DS} > V_P$, drain current ($I_D$) constant with respect to drain to source voltage ($V_{DS}$) is known as the saturation region.

## *5.4.5.4 Construction and Working of p-Channel E-MOSFET*

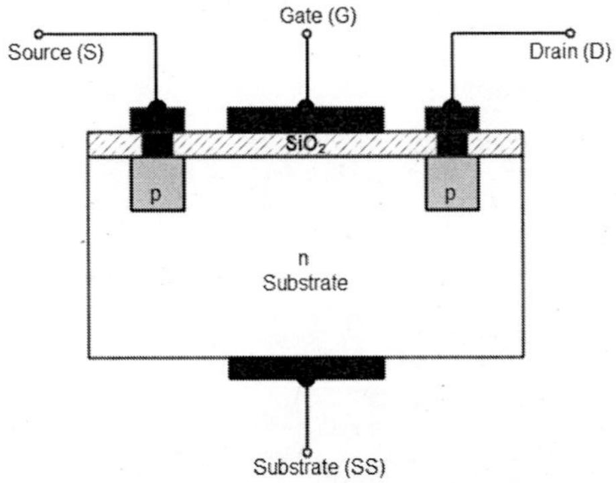


**Fig. 5.32:** P-channel Enhancement MOSFET

The basic construction of p-channel enhancement MOSFET is depicted in Fig. 5.32. The construction of p-channel E-MOSFET is exactly opposite of n-channel E-MOSFET. The source and drain are made with highly doped p-regions and the substrate is made with lightly doped n-regions. Here channel region is absent, that is no physical channel is present in between source and drain region. Here only the negative gate voltage operation is possible. Remaining all operations are same as the n-channel E-MOSFET.

## *5.4.5.5 Drain and Transfer Characteristics of p-Channel E-MOSFET*

The transfer characteristics of P-channel enhancement MOSFET is shown in Fig. 5.33 (a). It is a mirror image of an n-channel E-MOSFET shown in Fig. 5.31 (a). The drain characteristics of p-channel enhancement MOSFET is shown in Fig. 5.33 (b). Compared to n-channel E-MOSFET, p-channel E-MOSFET drain characteristics gate to source voltage ($V_{GS}$) polarity is changed.

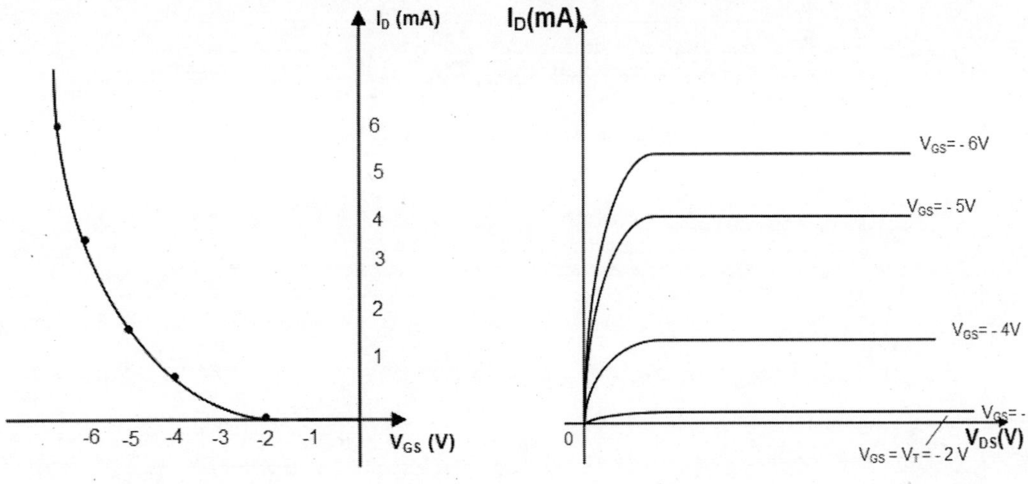


(a) (b)

**Fig. 5.33:** (a) Transfer Characteristics, (b) Drain Characteristics of P-channel E-MOSFET

### 5.4.6 Comparison of D-MOSFET and E-MOSFET

The comparison of D-MOSFET and E-MOSFET depends on symbol and characteristics are summarized in Table 5.3.

**Table 5.3:** Comparison of D-MOSFET and E-MOSFET

| S.No | Parameter | D-MOSFET | E-MOSFET |
|---|---|---|---|
| 1 | Symbols | 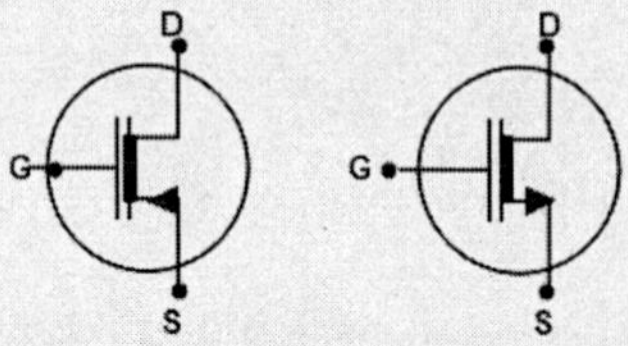  n-channel p-channel | 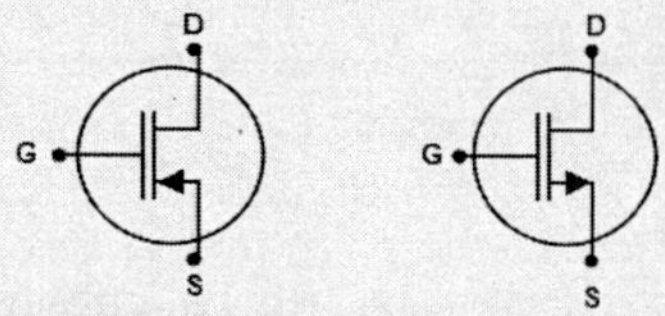  n-channel p-channel |
| 2 | Channel | In between source and drain region physical channel is present | Physical channel is absent in its structure between source and drain region |
| 3 | Operating Modes | Operated in both depletion mode and enhancement mode | Operated only in enhancement mode |
| 4 | Drain Characteristics | 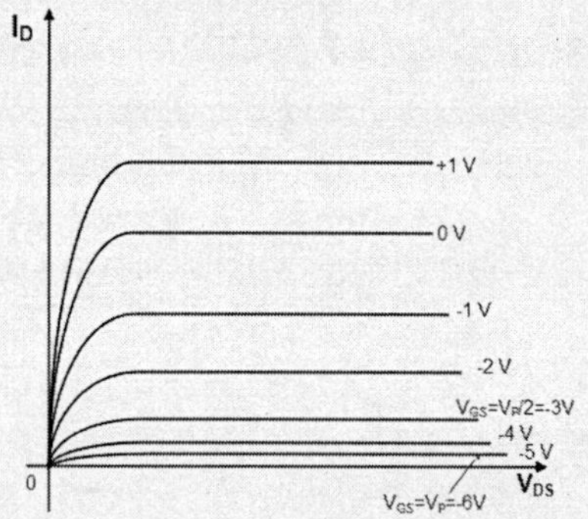  | 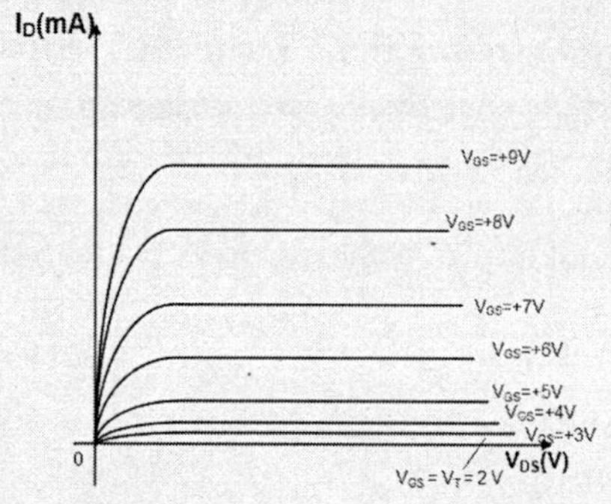  |
| 5 | Transfer Characteristics | 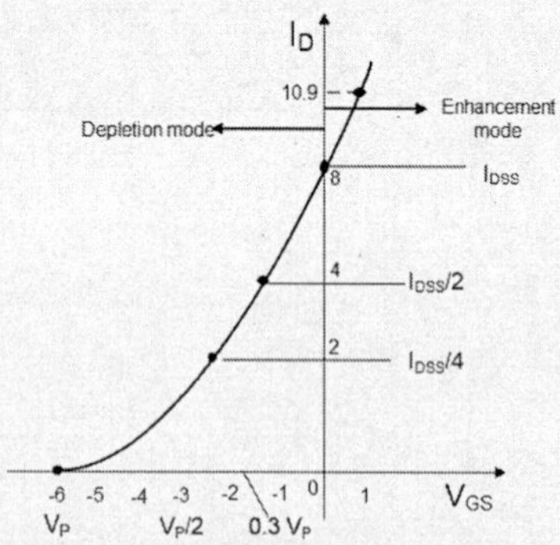  | 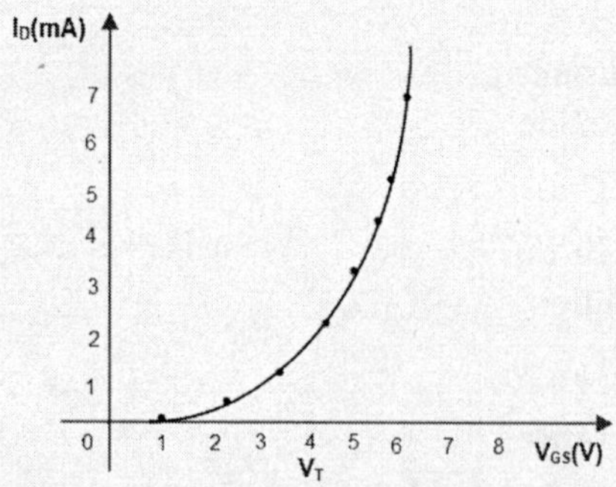  |
| 6 | Current Flow | At VGS = 0V, if drain to source voltage applied, drain current flows | At VGS = 0V, no current flows. If VGS > VT, then only drain current flows |

### 5.4.7 Comparison of JFET and MOSFET

The comparison of JFET and MOSFET depends on symbol and characteristics are summarized in Table 5.4.

**Table 5.4:** Comparison between JFET and MOSFET

| S.No | JFET | MOSFET |
|---|---|---|
| 1. | The input resistance is in the order of $10^9\Omega$. Since, there is no insulating layer between gate and the conducting channel. | The input resistance is very high in the order of $10^{13}\Omega$ due to the presence of insulating layer between gate and conducting channel. |
| 2. | The gate leakage current 0.1 to 10mA. | The gate leakage current is in the order of 0.1 to 10pA. |
| 3. | The drain resistance is in the order of 0.1 to $1M\Omega$. | The drain resistance is in the order of 1 to $50K\Omega$. |
| 4. | Electric field across the reverse biased PN junction controls the conductivity of the channel. | Electric field across the insulating layer controls the conductivity of the channel. |
| 5. | JEFT operates only in depletion mode. | The MOSFET operates both in enhancement mode and depletion mode. |
| 6. | $V_{GS}$ for an n-channel JFET cannot be allowed to go positive since that would forward bias the gate source p-n junction and cause a large gate current to flow. | Gate source voltage of a depletion mode MOSFET can be negative or positive. |
| 7. | Fabrication process is much more complex than that of MOSFET. | One can easily fabricate MOSFET. Thus, it is very much more in use. |

## 5.5 I-V CHARACTERISTICS OF MOSFET

MOSFET is operated in three operating conditions, such as cut off region, active region and saturation region.

The MOS transistor can be controlled by the applied voltage. So, it is modelled as voltage-controlled device.$I_{ds}$ is the important parameter that determines the speed of the switch so the value of $I_{ds}$ depends on distance between the source and drain, channel width, applied voltage on the gate, thickness of the gate oxide layer and the carrier mobility.

i. **Cut-off (Subthreshold Voltage)**

In this region $V_{GS} < V_t$ and $I_{ds} = 0$. Hence the channel is not created. The transistor is in OFF condition. There is no conduction between source and drain. The current will be zero.

$$I_{ds} = 0V \quad (5.34)$$

ii. **Linear or Non-Saturation Region**

In this region of operation $V_{gs} > V_t$ and $V_{ds} < V_{gs} - V_t$. The value of $I_{ds}$ depends on two factors that is number of charges in the channel and velocity of the charge.

The charge under the channel is defined by,

$$Q_{channel} = C_{ox}.\ W.L.V$$

$$V = V_{GS} - V_t = \left[V_{GS} - \frac{V_{ds}}{2}\right] - V_t \tag{5.35}$$

The velocity is direct proportion to the applied field between source and drain,

$$E = V_{ds} / L$$

The velocity of the charge carrier is given by, $V = \mu.\ E$ (5.36)

$$Q_{channel} = C_{ox}.\ W.L.\mu.\ \frac{V_{ds}}{L} \tag{5.37}$$

Time of the carriers to pass the channel, $t = L/V$

$$I_{ds} = \frac{Q_{channel}}{t} \tag{5.38}$$

$$= \frac{\mu C_{ox} W}{L\ .}\left[(V_{GS} - V_t) - \frac{V_{ds}}{2}\right]V_{ds}$$

$$I_{ds} = KV_{ds}\left[(V_{GS} - V_t) - \frac{V_{ds}}{2}\right] \tag{5.39}$$

where K is the conductance parameters, $K = \mu.C_{ox}.\frac{W}{L}$

iii. **Saturation Region**

In saturation region of operation $V_{GS} > V_t$, $V_{ds} > V_{GS} - V_t$. The channel pinch-off near the drain. The drain voltage no longer increases, the current is,

$$I_{ds} = KV_{ds}\left[(V_{GS} - V_t) - \frac{V_{ds}}{2}\right] \tag{5.40}$$

Substitute $V_{ds} = V_{GS} - V_t$,

$$I_{ds} = K\,[V_{GS} - V_t].\left[(V_{GS} - V_t) - \left(\frac{(V_{GS} - V_t)}{2}\right)\right] \tag{5.41}$$

$$I_{ds} = \frac{K}{2}(V_{GS} - V_t)^2 \tag{5.42}$$

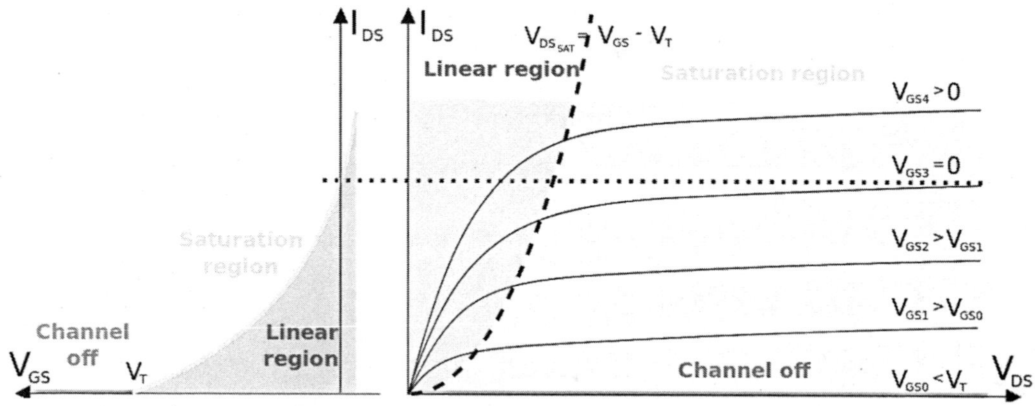


**Fig. 5.34:** I-V Characteristics of n-MOSFET

## 5.6 SMALL SIGNAL MODEL OF MOS TRANSISTOR

In order to analyse a transistor circuit, one needs a mathematical model or equivalent circuit of the transistor. One of the most useful models is the small-signal equivalent circuit. This model is used to determine the small signal voltage gain, input and output impedance, and other parameters of the amplifier circuit.

The small signal model of a MOS Transistor consists of two parts:

i. The small signal model parameters

ii. The AC equivalent circuit

To derive the small signal model of a MOS Transistor, the DC operating point of the MOS Transistor must be calculated first. Then, the AC equivalent circuit is derived by linearizing the MOS Transistor around the DC operating point. Finally, the small signal model parameters are calculated using the AC equivalent circuit. Fig. 5.35 (a) shows the n-channel, enhancement mode MOS Transistor common-source circuit. The time varying sinusoidal input signal along with the d.c source is applied on the gate. Fig. 5.35 (b) shows the MOS Transistor characteristics, d.c load line and Q point. MOS Transistor is biased in saturation region to get the output voltage as a linear function of input voltage. The total gate voltage is the sum of $V_{GSQ}$ and $V_i$. The time varying signal source generates a time-varying component of the gate to source voltage.

The instantaneous gate to source voltage is given by,

$$V_{GS} = V_{GSQ} + V_i = V_{GSQ} + V_{gs} \tag{5.43}$$

where $V_{GSQ}$ is the d.c component and $V_{GS}$ is the a.c component.

The drain current is given by,

$$i_D = K(V_{GS} - V_T)_2 \tag{5.44}$$

Substitute the equation $V_{GS}$ in equation 5.44 we get,

$$i_D = K(V_{GSQ} + V_{gs} - V_T)^2 = K[(V_{GSQ} - V_T) + V_{gs}]_2 \tag{5.45}$$

$$i_D = K(V_{GS} - V_T)^2 + 2K(V_{GSQ} - V_T)\, V_{gs} + K\, V_{gs}^2 \tag{5.46}$$

The first term in equation 5.46 represents the d.c or quiescent drain current $I_{DQ}$, the second term represents the time-varying drain current component and the third term is proportional to the square of the signal voltage.

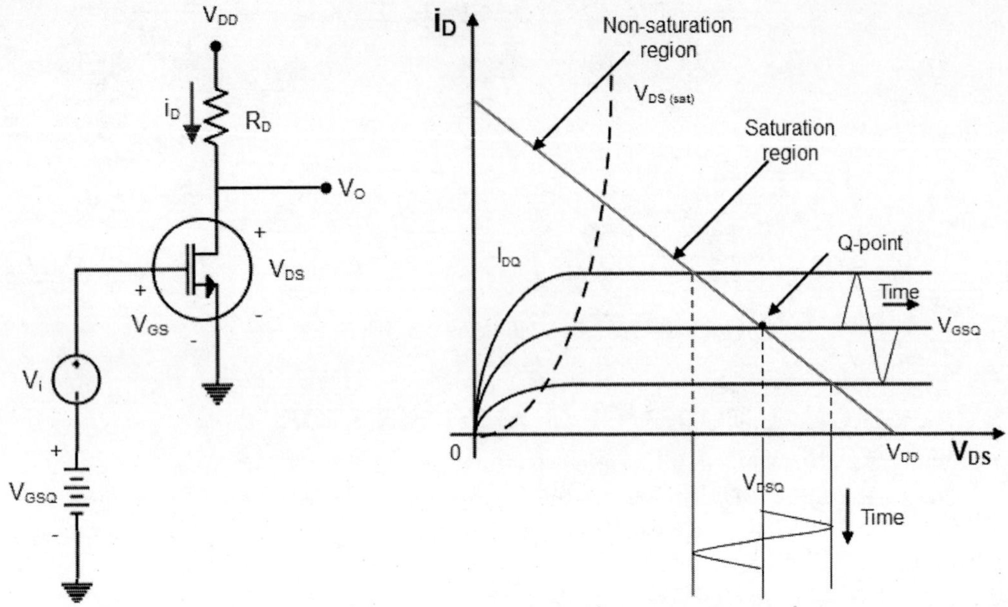


(a) (b)

**Fig. 5.35:** (a) NMOS Common source circuit with time varying signal, (b) Common source characteristics d.c load line

For a sinusoidal input signal, the squared term produces undesirable harmonics in the output signal. In order to minimize the harmonics, we must have,

$$V_{gs} \ll 2(V_{GSQ} - V_T) \tag{5.47}$$

Neglecting the term $V_{gs}^2$ term, we can write the equation 5.46,

$$i_D = I_{DQ} + i_d \tag{5.48}$$

where, $I_{DQ} = K(V_{GSQ} - V_T)^2$

$$i_d = 2K(V_{GSQ} - V_T)\, V_{gs} \tag{5.49}$$

The small signal current is related to the small signal gate to source voltage by the transconductance $g_m$. The relation is written by,

$$g_m = \frac{\partial id}{\partial V_{gs}} = 2K(V_{GSQ} - V_T) \tag{5.50}$$

The transconductance can also be written as, $g_m = 2\sqrt{KI_{DQ}}$

The above equation shows that, the transconductance is directly proportional to the conductance parameter K. Therefore, increasing the width of the MOSFET increases the transconductance.

## *Small Signal Equivalent Circuit*

From the Fig. 5.36, the output voltage is,

$$V_{DS} = V_o = V_{DD} - i_D R_D \quad (5.51)$$

By using the equation 5.48 we obtain,

$$V_o = V_{DD} - (I_{DQ} + i_d)\, R_D = (V_{DD} - I_{DQ}R_D) - i_d R_D \quad (5.52)$$

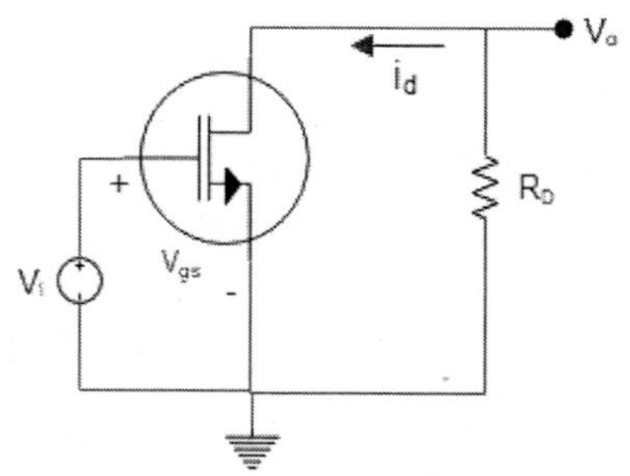


**Fig. 5.36:** A.C equivalent circuit of Common-Source Amplifier with NMOS transistor

The output voltage is also a combination of d.c and a.c values. The time varying output signal is written as,

$$V_o = V_{ds} = -\, i_d R_D \quad (5.53)$$

The Fig. 5.36 shows the a.c equivalent circuit. Here, the d.c sources are made zero. From this equivalent circuit, we can draw a small signal equivalent circuit for the MOSFET. Fig. 5.37 represents the small signal low frequency a.c equivalent circuit for n-channel MOSFET.

We know that the current $i_D = K[(V_{GS} - V_T)_2\,(1 + \lambda v_{DS})]$ (5.54)

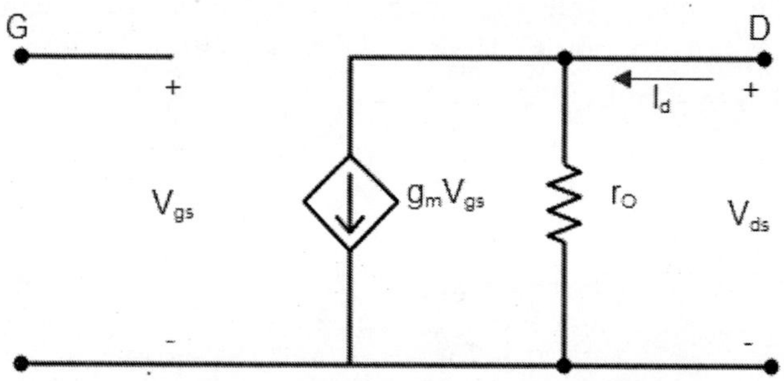


**Fig. 5.37:** Expanded small signal equivalent circuit, including output resistance for NMOS transistor

Here λ is the channel length modulation parameter. Considering the channel and modulation effects in this model, an output resistance is defined as,

$$r_0 = \left(\frac{\partial i_d}{\partial v_{DS}}\right)^{-1}\Bigg|_{v_{GS}=V_{GSQ}=\text{const}} \quad (5.55)$$

$$r_0 = \left[\lambda K\left(V_{GSQ} - V_T\right)\right]^{-1} \cong \left[\lambda I_{DQ}\right]^{-1} \quad (5.56)$$

Since the current source in the small-signal model is $g_m V_{gs}$ and the output resistance is $r_o$, finding the transconductance ($g_m$) will allow us to determine the amount of current flowing through the circuit. The variation of the drain-source voltage to the current is the output resistance, or $r_o$. The Fig. 5.38 shows the small signal equivalent circuit of the common-source circuit shown in Fig. 5.35.

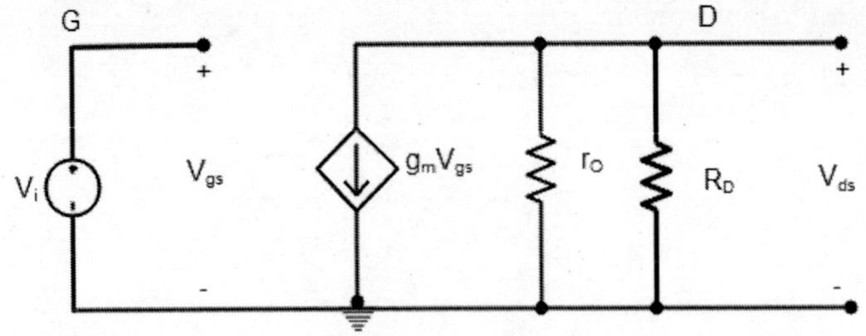


**Fig. 5.38:** Small Signal equivalent circuit of Common-Source with NMOS transistor

## APPLICATIONS

The Most significant invention of the 20th Century. What characteristics distinguish an innovation that actually changes the world? The jet engine, the aeroplane, television, the internet, the computer, and antibiotics are all deserving contenders for the most important invention of the 20th century, but unless you live like a hermit in the 15th century or are a member of an undiscovered tribe, there is one that affects almost everyone on the planet on a daily basis. Though few have seen them directly, they are now the most widely manufactured device in history.

The Regency TR-1, the first transistor radio in history, cost \$49.95 in 1954 (equal to \$507 in 2021) and used four Texas Instruments NPN transistors. Today's 512GB SD cards cost around \$30 and can hold over a trillion transistors. An estimated 13 sextillion, or 13 billion trillion, of them have been produced since 1960, and that number is continuing to expand at an accelerating rate as of 2018. Nothing else created by humans has undergone the rate of change the transistor has in little over 60 years, which has allowed for the development of other technologies and discoveries that have had a profound impact on society outside the realm of electronics.

## UNIT SUMMARY

- Work function of a metal, $\phi_m$- It is energy required to remove an electron at fermi level to the vacuum level outside the metal
- Electron affinity–It is the energy difference between reference vacuum level and the conduction band energy of a semiconductor

$$\chi = E\_0 - E\_C$$

- Work function of a semiconductor - $\phi_s = \chi + (E_C + E_F)$
- The energy difference, $\phi_n = (E_C - E_F)$
- The space charge region width,

$$W = x_d = \sqrt{2\varepsilon_s V_{bi}})/eN_d$$

- Flat band condition, $V_g = 0$

- Accumulation, $V_g < 0$
- Depletion, $V_g > 0$
- Inversion–with large positive bias
- Flat band condition– $\phi_s = 0$
- Inversion, $\phi_s = 2\phi_F$
- Biasing conditions,
  Accumulation– $\phi_s < 0$
  Depletion– $0 < \phi_s < \phi_F$
  Inversion– $\phi_s > \phi_F$
- Maximum depletion width,

$$W_{max} = \sqrt{\frac{2\varepsilon_s\varepsilon_0}{qN_a}}$$

- Capacitance Accumulation - $C = C_{ox} = \varepsilon_{ox}/t_{ox}$
- Depletion– $C = C_{dep}$
- Inversion – $C = C_{ox}$

## EXERCISES

### Multiple Choice Questions

5.1 ________ by the use of an electric field, a semiconductor's charge carrier's shape and resulting conductivity are controlled

a. FET b. BJT
c. MOSFET d. None

5.2 The FET utilises an ___________

a. Voltage field effect b. Electric field effect
c. Both a& b d. None

5.3 FET is frequently employed as an ____________

a. Oscillator b. Switch
c. Amplifier d. None

5.4 A p-type and n-type silicon bar with ________PN junctions at the sides makes up a FET

a. 1 b. 2
c. 3 d. 4

5.5 FET is a ________terminal using an active semiconductor, where the input voltage creates an electric field that controls the output current

a. 3 b. 2
c. 1 d. 4

5.6 The ________ applied to the gate can be used to regulate the amount of drain current from the source

a. Current b. Potential
c. Both a& b d. None

5.7 The characteristics of Junction Field Effect Transistors is________

a. High frequency switching
b. Slow switching
c. Fast switching
d. None

5.8 By maintaining the drain to source voltage at, the transfer characteristic between the gate voltage and the drain current is drawn ___________

a. Gain
b. Threshold voltage
c. Switch
d. Pinch-off voltage

5.9 When the gate terminal voltage is _________ and voltage VDS is placed between the drain and source terminals, two pn-junctions on either side form depletion layers

a. Zero
b. Positive
c. Negative
d. None

5.10 In ____________MOSFET, the channel is created after being initially non-existent

a. Depletion
b. Enhancement
c. Both a& b
d. None

5.11 In ____________MOSFET, as the MOSFET is being built, the channel is permanently created by doping

a. Both b & c
b. Enhancement
c. Depletion
d. None

5.12 Depletion type MOSFET are normally _______ type switches

a. OFF
b. Ground
c. Terminal
d. ON

5.13 In an enhancement-mode MOSFET, the electrostatic field created by the application of a gate voltage __________ the conductivity of the channel

a. Enhances
b. Deplete
c. Both a & b
d. None

5.14 In depletion mode, the electrostatic field created by the application of a gate voltage __________ the conductivity of the channel

a. Enhances
b. Deplete
c. Both a & b
d. None

5.15 An insulator, a metal electrode known as a metal electrode, and a semiconductor body or substrate make up a MOS capacitor ______

a. Source
b. Drain
c. Gate
d. None

5.16 An ___________has a boron doping-concentration of $10^{15}$ $cm^{-3}$ in the substrate

a. High MOS capacitor
b. Low MOS capacitor
c. Linear MOS capacitor
d. Ideal MOS capacitor

5.17 The MOS capacitor is under ________ inversion with VG = 2V

a. Strong
b. Weak
c. Both a & b
d. None

5.18 When no external voltage is provided to the MOS capacitor, the Fermi level of the metal and semiconductor are at equilibrium ________ level

a. Not same  b. Same
c. High  d. Low

5.19 $V_S$ in a small signal model is placed between ______ and ________ terminal

a. Gate and drain  b. Drain and source
c. Gate and source  d. None

5.20 There is a standard analysis technique in electronics engineering known as ____________ that employs linear equations to predict how nonlinear electronic circuits would behave

a. All  b. High signal model
c. Ebers-Moll Model  d. Small signal model

| Answer of Multiple-Choice Questions | | | | |
|---|---|---|---|---|
| 5.1 **a** | 5.2 **b** | 5.3 **c** | 5.4 **d** | 5.5 **a** |
| 5.6 **b** | 5.7 **c** | 5.8 **d** | 5.9 **a** | 5.10 **b** |
| 5.11 **c** | 5.12 **d** | 5.13 **a** | 5.14 **b** | 5.15 **c** |
| 5.16 **d** | 5.17 **a** | 5.18 **b** | 5.19 **c** | 5.20 **d** |

## SHORT AND LONG ANSWER TYPE QUESTIONS

### Category I

5.1 Define FET.

5.2 Differentiate FET and BJT.

5.3 List out the types of FET.

5.4 Explain the features of FET.

5.5 Sketch the I-V characteristics of MOSFET.

5.6 Differentiate n-channel and p-channel FET.

5.7 Explain the term MOSFET.

5.8 Differentiate depletion and enhancement type MOSFET.

5.9 Draw the symbol of MOSFET.

5.10 Sketch the drain and transfer characteristics of n-channel depletion type MOSFET.

5.11 Define MOS capacitor.

5.12 Explain the band diagram of MOS capacitor.

5.13 Draw the band diagram of accumulation, depletion and inversion mode.

5.14 Express the three operating conditions capacitance equation.

5.15 Sketch C-V characteristics curve

5.16 Draw the symbol of FET.

5.17 Give three different operating conditions of MOSFET.

5.18 Draw the symbol of MOSFET.

5.19 Discuss about small-signal model of MOSFET.

5.20 Draw the small signal diagram of MOSFET.

## Category II

5.1 Explain in brief about construction and working principle of FET.

5.2 Explain in detail about enhancement type MOSFET

5.3 Discuss about depletion mode MOSFET.

5.4 Illustrate the working principle of MOSFET with I-V characteristics.

5.5 Explain about MOS capacitor and operating modes.

5.6 Illustrate about small signal mode of MOSFET.

## *Numerical Problems*

5.1 For the circuit shown, calculate drain current.
**(Ans: 6.12mA)**

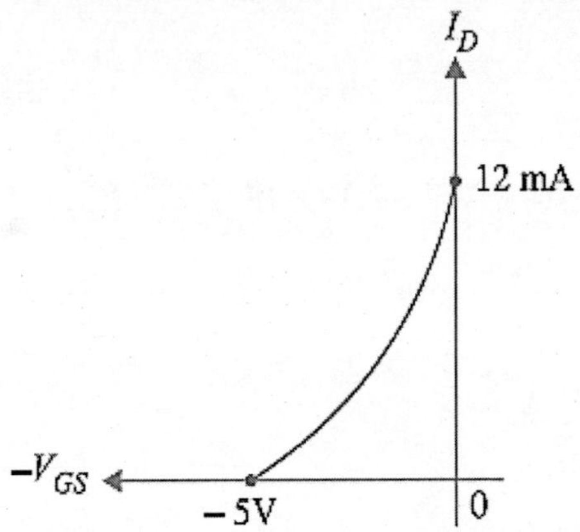


5.2 Consider the JFET having following properties: $I_{DSS}$ = 24mA; $V_{GS}$ (off) = –2V, $V_{GS}$ = – 4.5V. calculate the value of drain current.
**(Ans: 37.5mA)**

5.3 For the diagram depicted in the following illustration. Identify the bare minimum VDD value necessary to place the device in the saturation region.
**(Ans: 10.72V)**

5.4 A JFET has a pinch-off voltage of –8V and a drain current of 2 mA when the gate-source voltage is -4V. What is the drain-source voltage of the JFET?
**(Ans: 1.6V)**

5.5 A JFET with a pinch-off voltage of –6V has a drain current of 3 mA when the gate-source voltage is -2V. What will be the drain current if the gate-source voltage is increased to -4V?
**(Ans: 2.67 mA)**

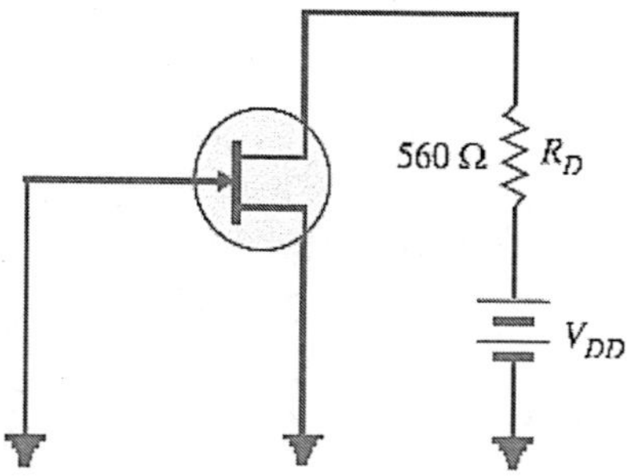


5.6 JFET has the value of gmo = 4000μs. Calculate the $g_m$ value at $V_{GS} = -3V$. Assume that $V_{GS}$ (off)=-8V.

**(Ans: 2500 $\mu s \mu s$ )**

5.7 The transfer characteristics of a JFET reveals that when $V_{GS} = -5V$, $I_D = 6.25mA$. find the value of source resistance.

**(Ans: 800Ω)**

5.8 Find the value of $R_s$ to self-bias a p-channel FET having $I_{Dss} = 20mA$,$V_{GS}$ (off)=15V and $V_{GS} = 5V$.

**(Ans: 561Ω)**

5.9 A JFET has a threshold voltage of -5V and a drain current of 1 mA when the gate-source voltage is -3V. What is the transconductance of the JFET?

**(Ans: 2.4 ms)**

5.10 A MOSFET has a threshold voltage of 2V and a gate-source voltage of 5V. The channel length is 1μm, the channel width is 20μm, and the gate oxide capacitance per unit area is 0.5μF/cm². The drain-source resistance is 500 Ω. What is the drain current if the drain-source voltage is 10V? Assume: μnCox = 100 μA/V²

**(Ans: -240mA)**

5.11 A MOSFET has a threshold voltage of 2V and a channel length of 1 μm. If the gate-source voltage is 1.5V and the drain-source voltage is 3V, determine the region of operation of the MOSFET and the drain current.

**(Ans: $I_D = 0A$, cut off region).**

5.12 A MOSFET has a threshold voltage of 3V and a gate-source voltage of 8V. The channel length is 2μm, the channel width is 40μm, and the gate oxide capacitance per unit area is 0.8μF/cm². The drain-source resistance is 1kΩ. What is the drain current if the drain-source voltage is 15V? Assume: μnCox = 120 μA/V².

**(Ans: 7.2mA)**

5.13 A MOSFET has a threshold voltage of 1.5V and a channel length of 2μm. If the gate-source voltage is 2V and the drain-source voltage is 5V, determine the region of operation of the MOSFET and the drain current. Assume μnCox = 150 μA/V².

**(Ans: Saturation region, $I_D = 562.5\mu A$)**

5.14 A JFET has a pinch-off voltage of -5V and a drain-source voltage of 10V. If the gate-source voltage is -2V and the drain current is 2 mA, determine the value of the drain-source resistance. Assume a transconductance parameter of 4 mA/V and a gate-source cutoff voltage of -2V.

**(Ans: 3.91KΩ)**

## KNOW MORE

Julius Edgar Lilienfeld was the first to patent the fundamental design of this type of transistor in 1925.

William Shockley, John Bardeen, and Walter Houser Brattain, scientists from Bell who were involved in the research that resulted in the discovery of the transistor effect, presented the design of a device resembling a MOS transistor. Due to the surface state issue, which is caused by traps on the semiconductor surface that keep electrons immobile, the structure failed to produce the desired results. The silicon wafer was unintentionally covered in a layer of silicon dioxide in 1955 by Carl Frosch and L. Derick.

**Julius Edgar Lilinenfeld**

Further investigation revealed that silicon dioxide might stop the diffusion of dopant into silicon wafers. Mohamed M. Atalla built on this work by demonstrating how silicon dioxide is particularly successful in resolving the issue of a crucial class of surface states. Atalla spent this time researching the surface characteristics of semiconductors and soon discovered strategies for facilitating the passage of electricity to a chipset's semiconducting layer. By developing silicon dioxide layers on top of silicon wafers, he was able to do this.

**Mohamed M. Atalla**

This procedure, which came to be known as surface passivation, made semiconductor technology more widely used. he founded Atalla Corporation, and filed a patent for a remote Personal Identification Number (PIN) security system. In 1973, he released the first hardware security module, the "Atalla Box", which encrypted PIN and ATM messages, and went on to secure the majority of the world's ATM transactions.

**Dawon Kahng**

## REFERENCES AND SUGGESTED READINGS

1. David.J.Griffiths, *Introduction to Quantum Mechanics,* Pearson Prentice Hall, 2005.
2. Chenming Calvin Hu, *Modern Semiconductor Devices for Integrated Circuits,* Pearson Prentice Hall, 2009.
3. S.O.KASAP, *Principle of Electronic Materials & Devices,* MC Graw Hill Education, 2018.
4. S.Datta, *Quantum Transport:Atom to Transistor,* Cambridge University Press, 2005.
5. Robert F. Pierret, *Semiconductor Device Fundamentals,* ADDISON-WESLEY Publishing Company, 1996.
6. Donald A. Neamen, *Semiconductor Physics and Devices Basic Principle,* MC Graw Hill Companies, 2012.
7. Ben G, Streetman, Sanjay Kumar Banerjee, *Solid State Electronic Devices,* Pearson Education Limited, 2016.
8. *https://physicsinmyview.com/2020/09/facts-about-quantum-physics.html*
9. *https://onlinecourses.nptel.ac.in/noc21_ee80/course*
10. *https://onlinecourses.nptel.ac.in/noc21_ee59/course*
11. *https://onlinecourses.nptel.ac.in/noc21_ee19/course*
12. https://en.wikipedia.org/wiki/BJT.

# 06 INTEGRATED CIRCUIT FABRICATION PROCESS

## UNIT SPECIFICS

*Through this unit we have discussed the following aspects:*

- *Historical transformation of semiconductor technology*
- *Fabrication steps and introduction about the crystal growth*
- *The focus of the epitaxy in IC fabrication*
- *Fundamentals of fabrication process*
- *Discuss about the steps to create silicon devices*
- *Problems including constraints and friction*

*The practical applications of the topics are discussed for generating further curiosity and creativity as well as improving problem solving capacity.*

*Besides giving a large number of multiple choice questions as well as questions of short and long answer types marked in two categories following the lower and higher order of Bloom's taxonomy, assignments through a number of numerical problems, a list of references and suggested readings are given in the unit. so that one can go through them for practice. It is important to note that for getting more information on various topics of interest some QR codes have been provided in different sections can be scanned for relevant supportive knowledge.*

*After the related practical, based on the content, there is a "Know More" section. This section has been carefully designed so that the supplementary information provided in this part becomes beneficial for the users of the book. This section mainly highlights the initial activity, examples of some interesting facts, history of the development of the subject focusing the salient observations and finding, timelines starting from the development of the concerned topics up to the recent time, applications of the subject matter for our day-to-day real life or/and industrial applications on variety of aspects, case study related to environmental, sustainability, social and ethical issues whichever applicable, and inquisitiveness and curiosity topics of the unit.*

## RATIONALE

*IC Technology in the expanding market where there are many chances for expansion. It is always demanded by fabrications. According to a recent poll, there are presently more integrated circuits than people in the UK, USA, India, and China. Microelectronic engineering has become a separate subject, as a result of the amazing developments in the creation and use of integrated circuit technology. The manufacturing process is often broken down into a series of unit procedures that are repeated to create the integrated circuit in this unit.*

*All significant phases in the manufacturing process are covered from both a theoretical and practical standpoint. This unit may be utilised easily in the course of integrated circuits. Additionally, this unit may be used as a reference for engineers and scientists, in the semiconductor sector.*

## PRE-REQUISITES

*Basics of BJT & MOSFET*

## UNIT OUTCOMES

*List of outcomes of this unit is as follows:*

*U6-O1: Describe the fabrication process flow*

*U6-O2: Describe the process of Wafer formation and Ion implantation*

*U6-O3: Explain the chemical vapour deposition process*

*U6-O4: Realize the integrated circuits in the semiconductor industry*

*U6-O5: Discuss the future semiconductor technology*

| Unit-6 Outcomes | EXPECTED MAPPING WITH COURSE OUTCOMES (1- Weak Correlation; 2- Medium Correlation; 3- Strong Correlation) | | | | | |
|---|---|---|---|---|---|---|
| | CO-1 | CO-2 | CO-3 | CO-4 | CO-5 | CO-6 |
| U6-O1 | 3 | 3 | 3 | - | 3 | 1 |
| U6-O2 | 1 | 1 | 2 | 2 | 1 | - |
| U6-O3 | 2 | 1 | 3 | 1 | 2 | 1 |
| U6-O4 | - | - | 3 | 1 | 2 | 2 |
| U6-O5 | 3 | 3 | 3 | - | 3 | 1 |

## 6.1 INTRODUCTION

An Integrated Circuit (IC) is a complicated design of electronic devices, including a large number of components. An IC is a collection of electronic devices, fabricated on a small chip of semiconductor material. The speciality of integrated circuits is to control the operation of devices with semiconductor material.

Due to its mix of qualities and affordability, Silicon (Si) is highly preferable for manufacturing. Germanium (Ge) and Gallium Arsenide (GaAs) are two less popular semiconductor elements. ICs are fabricated from silicon due to its vast range of advantages over other materials. This Silicon is used in many applications. Some of them are listed:

- Silicon is made from silica. Silica is a raw ingredient used in the production of concrete.
- Because of its hardness and scratch resistance, Silica is used in varnishes.
- The glass industry, including the production of quartz glass, utilizes silica as a fundamental raw material.
- When tyres are made, amorphous silica is used as filler in the rubber. This aids in lowering the vehicle's fuel usage.
- ICs are mostly square or rectangular flat plate chips, which is approximately 0.50 mm thickness and ranging from 5.0 to 25.0 mm all sides.

## 6.1.1 Moore's Law

In 1965, Gordon Moore, co-founder of Fairchild Semiconductor and Intel found the plotting between the number of transistors on chip vs. year in semi-log scale, known as Moore's Law. Moore's Law states that the number of transistors per chip grows exponentially or doubles in every 18 months. The Moore's law plot is shown in Fig. 6.1.

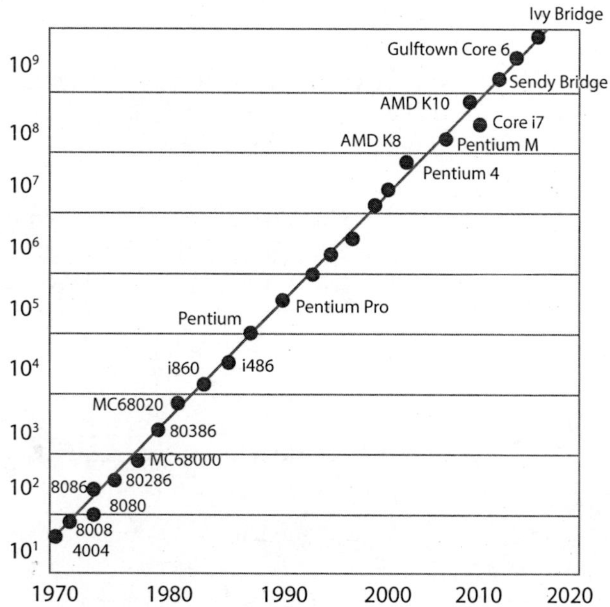


**Fig. 6.1:** Moore's Law

## 6.1.2 Classification of ICs

ICs can be classified into different types based on different criteria. ICs can be classified based on fabrication method/technique, integration scale/level and application. The classification of IC is depicted in Fig. 6.2.

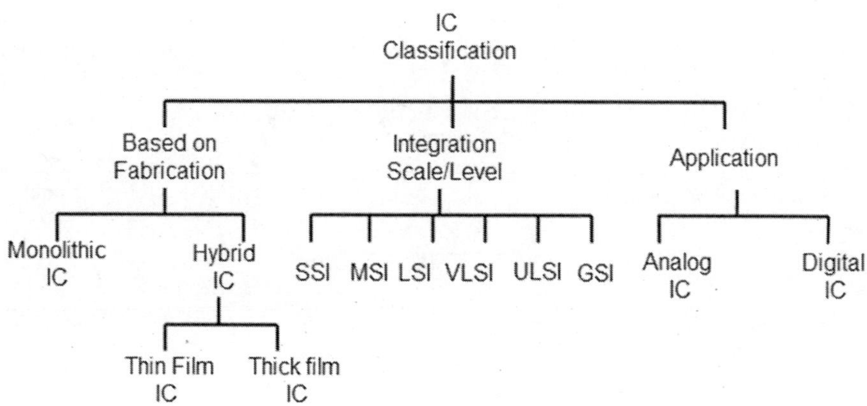


**Fig. 6.2:** Classification of ICs

a. **Classification Based on Fabrication**

Integrated circuits can be classified as monolithic IC and hybrid IC based on fabrication method / technique.

1. **Monolithic IC (MIC)** – In monolithic ICs, the entire ICs are fabricated on a single chip of a semiconductor.
2. **Hybrid IC (HIC)** – In hybrid ICs, the number of individual chips are fabricated by interconnection. It can be a combination of monolithic and film techniques. Hybrid ICs are further classified as thin film IC and thick film IC.
   i. **Thin Film IC** – By depositing layers of conducting material on the surface of a glass or ceramic foundation, thin film integrated circuits are formed. Resistors and conductors can be fabricated by using different materials and by controlling the width and thickness of the films.
   ii. **Thick Film IC** – Thick film ICs are integrated circuits made out of resistors, transistors, crystal diodes, and semiconductor integrated circuits that are created on insulating materials like ceramic sheets or glass. It is utilized for switching TV or audio system power supply circuits.

b. **Classification Based on Integration Scale / Level**

Integrated circuits can be classified as SSI, MSI, LSI, VLSI, ULSI and GSI based on their density or the number of components they incorporate.

1. **Small Scale Integration (SSI)** – In SSI, the number of transistors in a package is less than 10. Examples: Logic gates (AND, OR, NOT, NAND, NOR).
2. **Medium Scale Integration (MSI)** – In MSI, the number of transistors in a package is more than 10 and less than 500. Examples: Flip flops, adders/counters, multiplexers and demultiplexers.
3. **Large Scale Integration (LSI)** – In LSI, the number of transistors in one package is more than 500 and less than 20000. Examples: Small memory chips and programmable logic devices.
4. **Very Large-Scale Integration (VLSI)** – In VLSI, the number of transistors in a package is more than 20000 and less than 100000. Examples: Large memory chips and complex programmable logic devices.
5. **Ultra Large-Scale Integration (ULSI)** – In ULSI, the number of transistors in one package is more than 100000 and less than 1000000. Examples: 8 and 16-bit microprocessors.
6. **Giga Scale Integration (GSI)** – In GSI, the number of transistors in a package is more than 1000000. Examples: Pentium IV processor.

The different IC types based on integration scale / level and its corresponding number of logic gates and transistors are tabulated in Table 6.1.

**Table 6.1:** Integration of IC Density

| Integration Scale / Level | | No. of Logic Gate used | No. of Transistor used |
|---|---|---|---|
| SSI | Small Scale Integration | 1-12 | 1-10 |
| MSI | Medium Scale Integration | 13-99 | 10-500 |
| LSI | Large Scale Integration | 100-9999 | 500-20000 |
| VLSI | Very Large-Scale Integration | 10,000-100,000 | 20000-1000000 |
| ULSI | Ultra Large-Scale Integration | 100,000-1,000,000 | 1000000-10000000 |
| GSI | Giga Scale Integration | >1,000,000 | >10000000 |

c. **Classification Based on application**

Integrated circuits can be classified as analog IC and digital IC based on its application.

1. **Analog IC** – In general, analog ICs are utilized for linear or amplification operations. Operational amplifiers, linear regulators, oscillators, active filters, and phase-locked loops are all designed using analog integrated circuits. When constructing analog integrated circuits, the semiconductor properties of power dissipation, gain, and resistance are of greater importance. Contiguous signals are handled by analog ICs.
2. **Digital IC** – Binary values (0 and 1) are discrete signals that are handled by digital integrated circuits. These circuits make use of flip flops, multiplexers, and digital logic gates. These circuits are more cost-effective and simpler to construct.

## 6.1.3 Advantages of Integrated Circuits

Integrated circuits are used in recent technology due to its advantages. Some of them are listed:

- ICs have compact size and less weight compared to discrete circuits.
- Due to its small size, IC has low power consumption and lower threshold power consumption.
- IC uses low-cost materials and ICs are fabricated as mass production. So, the production cost is lower.
- It has higher reliability.
- It plays a crucial role in making it easier to connect a large number of devices and components.
- Enhances device performance even in the high-frequency range.

## 6.1.4 Disadvantages of Integrated Circuits

Even though IC has many advantages, there are also some disadvantages. They are listed below:

- The fabrication of IC is limited in its range.
- Inductors cannot be produced using IC due to their large size.
- ICs cannot be used to make transformers.

## 6.1.5 Process Flow of IC

Silicon processing involves turning sand into very pure silicon, which is subsequently formed inside the wafer. Integrated Circuit manufacturing involves stages such as include, change and neglect of tiny layers in specific places for creation of electronic devices. The portions of the wafer surface that need to be processed are defined using lithography. IC packaging involves testing the wafer, cutting it into individual chips, and encasing the chips in the proper packaging. The process of making IC is shown in Fig. 6.3.

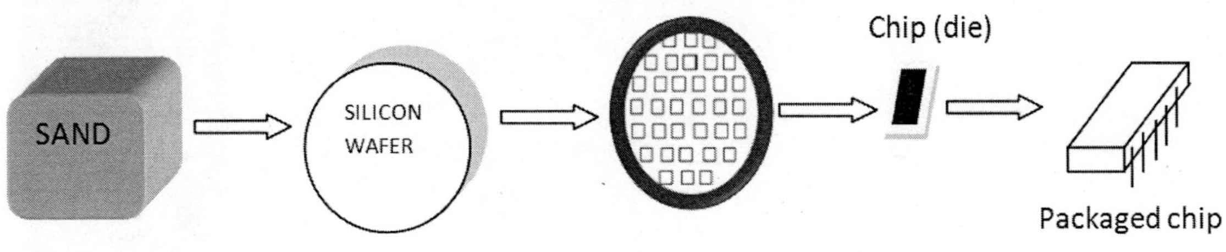


**Fig. 6.3:** Process of ICs

## 6.2 CMOS Fabrication

Complementary Metal Oxide Semiconductors, also known as CMOS, are made up of both NMOS and PMOS transistors. The terms NMOS and PMOS refer to N-type and P-type metal oxide semiconductors, respectively. Pentavalent impurities of the N-type and trivalent impurities of the P-type variety are both types of doped impurities on semiconductors. The transistors' three terminals are labelled as Gate (G), Source (S), and Drain (D). The D and S terminals are subjected to p-type/n-type doping.

Amplification devices, switching circuits, logic circuits, integrated circuit chips, microprocessors, and other devices are using CMOS transistors. Because of its low power dissipation and low operating currents, CMOS is significant in semiconductor technology. Compared to the production of Field Effect Transistors and Bipolar Junction Transistors, it needs fewer processing. There are three types in CMOS fabrication. They are listed as follows:

1. N-Tub Fabrication
2. P-Tub Fabrication
3. Twin-Tub Fabrication

In N-Tub fabrication, the substrate is of the p-type. In P-Tub fabrication, the substrate is of the n-type. In both N-Tub & P-Tub fabrication method, remaining steps follows monolithic IC fabrication technology. In Twin-Tub fabrication, a combination of p-well and n-well processes created on the same substrate. In N-Tub fabrication, the majority carriers for current flow are electrons. So, N-Tub fabrication method is mostly used.

## 6.3 FABRICATION PROCESS OF MONOLITHIC IC

The construction of an industrial product is known as fabrication. It can also be described as a group of procedures used in the production of electronic devices. Large electronic components can be created using the smaller components that were created during the fabrication process. The Greek words "Monos" and "Lithos," which mean "single" and "stone," are where the word "monolithic" originates. Thus, the term "monolithic integrated circuit" refers to a single crystal or stone.

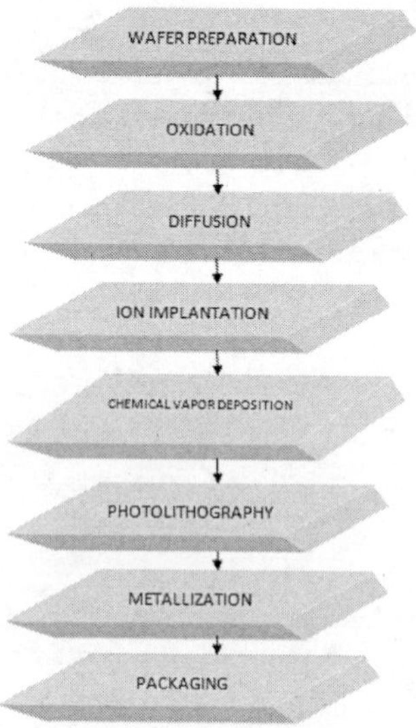


Scan to know more

**Fig. 6.4:** IC Fabrication Process

A single silicon chip serving as the single crystal is the semiconductor material on top of which all the passive and active parts are developed. As shown in Fig. 6.4, the process of making a monolithic integrated circuit (IC) consists of the following steps:

i. Wafer preparation
ii. Oxidation (Deposition of an oxide layer or protective layer)
iii. Diffusion (Introduction of impurities (n-type or p-type))
iv. Ion Implantation (Implantation of ions onto the crystal's surface)
v. Chemical Vapour Deposition (CVD to produce thin films)
vi. Photolithography (the use of UV light to create a mask pattern)
vii. Metallization (Deposition of the metal for interconnections)
viii. Packaging

Let's consider each fabrication step in detail in the upcoming sections.

## 6.3.1 Wafer Preparation

A wafer is made of semiconductors, particularly crystalline silicon. On a thin semiconductor disc, large numbers of integrated circuits are produced at the same time. The wafer preparation can be divided into three steps. They are,

- Production of electronic grade Silicon
- Crystal growing
- Shaping of silicon into wafers

### *6.3.1.1 Production of Electronic Grade Silicon*

Pure silicon is also known as Metallurgical Grade Silicon (MGS). In MGS, impurities are in the order of parts per million. But it is unsuitable for electronic device fabrication. MGS is further purified and its impurities are made in the order of parts per billion. This highly purified MGS is called as Electronic Grade Silicon (EGS) or semiconductor grade silicon (SGS), which is used for microelectronic device applications.

### *6.3.1.2 Crystal Growing*

The electronic grade silicon is used as starting material for crystal growth. The most popular method for growing silicon crystals is Czochralski growth. The float-zone procedure is another silicon technology that produces less contamination than the Czochralski method. Float-zone crystals are typically utilized in high-power, high-voltage devices that demand high resistivity materials. The float-zone approach produces small diameter wafers (<150 mm), whereas the Czochralski technique produces large diameter wafers (>350 mm). To form the n-type or p-type material, more impurities are added to the molten state in a precise quantity.

### *6.3.1.3 Shaping of Silicon into Wafers*

The silicon shaping produces perfect flat wafers as shown in Fig. 6.5. This process involves three steps. They are,

- Ingot (boule) preparation
- Wafer slicing
- Wafer preparation

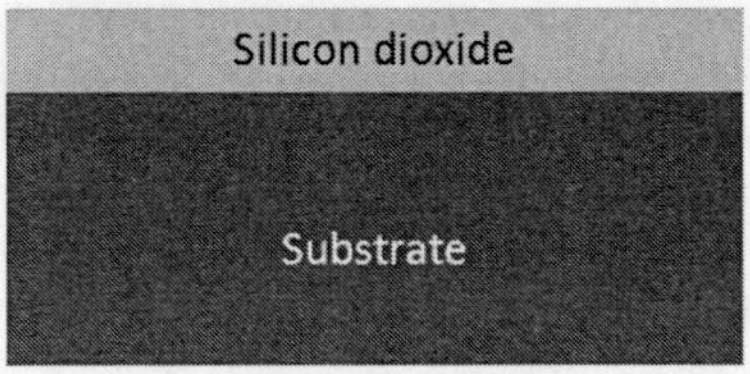


**Fig. 6.5:** Wafer

The ingot resistivity is checked to confirm the dopant concentration. First, the ingot (boule) ends are cut-off. To make the boule into perfect cylinder, the cylindrical grinding method is used. By using this method, the entire ingot surface has uniform diameter. After grinding, the ingot surface orientation is checked.

In the context of ingots used for semiconductor manufacturing, specifically for silicon wafers, the terms "primary flat" and "secondary flat" refer to specific orientations or markings on the ingot.

- Primary Flat–The primary flat is a flat surface or notch that is intentionally cut or ground onto the ingot during its manufacturing process.
- Secondary Flat–This is employed to identify the wafer, the type of dopant, and the orientation.

To prepare the wafer material for further production, it must be sliced, shaped and polished. During handling, the rims of the wafer are rounded fine by contour grinding wheels to shorten chipping. The regions are created using the planar method. An operation of polishing is carried out to produce highly smooth surfaces for subsequent processes of photolithography. At last, the wafer is chemically stained to abolish organic films and residues.

## 6.3.2 Oxidation

Process of introducing oxygen in Silicon wafer is called as Oxidation. The silicon wafer is protected against numerous contaminants by the deposition of silicon dioxide on it. $SiO_2$ is applicable to segregate device from each circuit. It can be applicable as a mask to stop the dopants of diffusion or ion implantation in Si wafer.

$SiO_2$ is one of the insulators, as inverse to Si. Silicon dioxide is created when silicon and oxygen interact in a semiconductor. High temperatures, up to 1250oC, are used for the oxidation process in furnaces.

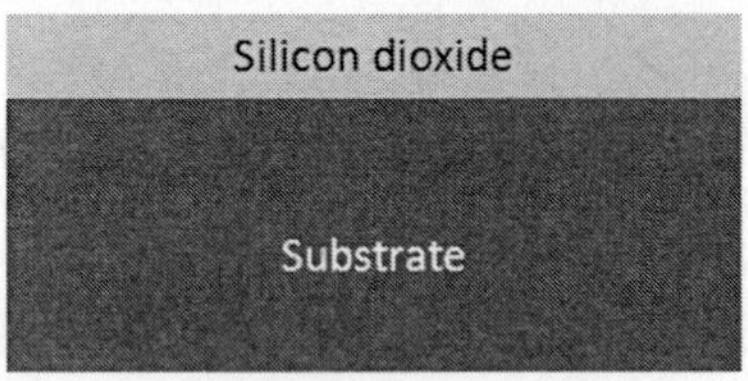


**Fig. 6.6:** Oxidation

The oxidation procedure includes depositing a thin layer of $SiO_2$ on silicon wafer surface as shown in Fig. 6.6. Si-wafers are exposed to a gas containing either $O_2$ or $H_2O$ or both, whiles being heated to a temperature between 950°C and 1150°C.

$$S_i + 2H_2O \rightarrow SiO_2 + 2H_2$$

Oxidation isolates one device from another while providing surface passivation. As the oxide layer is grown at a high temperature, this oxidation process is often referred to as thermal oxidation. Typically, the thickness ranges from 0.02 to 2 μm. The illustration of the diameter of $SiO_2$ is shown in Fig. 6.7. The oxidation can be classified into two types. They are dry oxidation and wet oxidation.

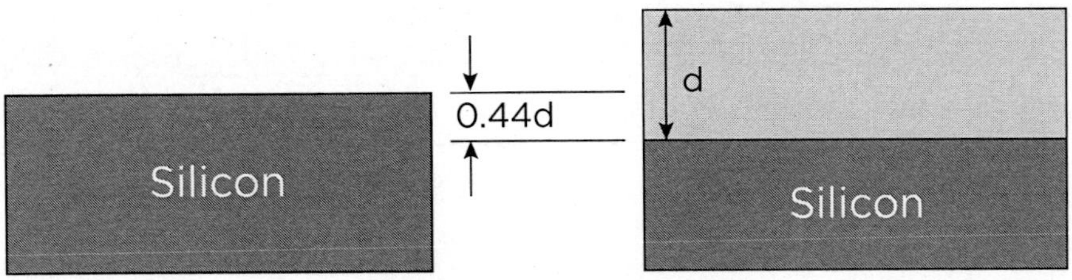


**Fig. 6.7:** Diameter of $SiO_2$

### 6.3.2.1 Dry Oxidation

The wafer is placed in a pure oxygen gas ($O_2$) atmosphere during the dry oxidation process. Between the solid silicon atoms (Si) on the wafer's surface and the incoming oxide gas, a chemical reaction occurs. In general, this is accomplished at temperature ranging from 1000°C to 1200°C. The procedure can be accomplished at a much lower temperature, about 900°C to produce a very thin and stable oxide layer. As a result, it is a relatively slow process. To get a desired thickness, it can be precisely adjusted. The chemical reaction at dry oxidation is,

$$Si + O_2 \rightarrow SiO_2$$

When compared to oxide films developed in a moist atmosphere, the films produced by this technique are of higher quality. As a result, they are in higher demand when high-quality oxides are required. Dry oxidation is typically employed to make thin films less than 110 nm thick or as a phase in the formation of thicker films. When forming thick oxides, widens the oxidation rate bottleneck. The second procedure is used to improve the thick oxide's quality. In this, the following features of dry oxidation are present:

1. Slow oxide formation
2. High density
3. High breakdown voltage

### 6.3.2.2 Wet Oxidation

Oxidation by water vapour is named as wet oxidation. The silicon wafer is engrossed in water vapour ($H_2O$) atmosphere during the wet oxidation process. Between the water vapour molecules and the solid silicon atoms (Si), a chemical reaction occurs. As a by-product of this oxidation process, hydrogen gas ($H_2$) is produced. On the other hand, 900°C to 1000°C temperature is used in this technique.

$$Si + 2H_2O \rightarrow SiO_2 + 2H_2$$

The following are the properties of wet thermal oxidation:

1. Very rapid development
2. Lower grade than dry oxides

The wet oxidation procedure has substantially higher oxidation rates than the dry oxidation process, by about 600 nm/h. Wet oxidation is utilised when thick oxides are required due to its quick development rate. Example: masking, insulation and passivation layers.

#### 6.3.2.3 Temperature Effects on Oxide Thickness

In both wet and dry oxidation process, increasing the temperature of the oxidation environment can greatly enhance the oxidation rate. Because oxidant diffusivity increases exponentially with temperature, oxidation rates should also increase. When thicker oxides (30 nm) are formed, the diffusivity of oxidants becomes the rate-limiting step.

#### 6.3.2.4 Impurities Effect on the Oxidation Rate

The oxidation rate is affected by various impurities such as,

1. $H_2O$
2. Na
3. III and V group elements
4. Halogens

The oxidation rate is also impacted by silicon damage. Any inadvertent moisture accelerates up dry oxidation because wet oxidation happens at a much faster rate than dry oxygen. High sodium concentrations affect the rate of oxidation by altering the bond structure of the oxide, which improves the oxygen molecule concentration and diffusion.

#### 6.3.2.5 Oxidation Techniques

To grow oxide layer in IC fabrication, chemical deposition and physical deposition techniques can be used. Various methods are available for oxide layer deposition techniques. They are listed as follows,

1. Chemical deposition
   1.1 Thermal oxidation
   1.2 CVD (Chemical Vapor Deposition)
   1.3 Epitaxy
   1.4 Electro position
2. Physical deposition
   2.1 PVD (Physical Vapor Deposition)
   2.2 Spin costing

#### 6.3.2.6 Roles of Oxidation

Oxidation plays a major role in integrated circuit fabrication. Some of the important points are listed as below:

- It acts as a mask for dopants during the diffusion and ion implantation procedure. It functions as a diffusion mask, allowing for selective implantation into silicon wafers. This is accomplished by etching a window into the oxide.
- Surface passivation is provided by the oxide layer. On the wafer surface, it generates a $SiO_2$ shield. It protects the joint from dampness.
- Device isolation is another aspect in oxidation. On the wafer's surface, it produces an insulating layer.

- As a MOS structural component (gate oxides), the active gate electrode in MOS devices is silicon dioxide.
- Oxidation Provides electrical isolation in multi-level metallization systems.
- Oxidation provides high thermal conductivity.
- Due to oxidation, conductors have no leakage between them.

## 6.3.3 Diffusion

Diffusion is a process used to introduce specific impurities or dopants into desired regions of a silicon wafer. The diffusion process typically involves two main steps: pre-deposition and drive-in diffusion. In the pre-deposition stage, a vapour containing the dopant is delivered at a high concentration on the silicon surface, at a temperature of around 1000°C. Silicon atoms displace from their lattice positions at a temperature of 1000°C, causing a high density of vacancies and rupturing the link between the surrounding atoms. The drive-in technique, which is employed in the second stage, is used to push the contaminants farther below the surface without introducing new impurities.

The diffusion of contaminants in the silicon chip is a crucial step in the production of integrated circuits as shown in Fig. 6.8. For this, a high-temperature furnace with a 20-inch-long temperature profile is used. Wafers are put into a quartz furnace tube that has resistance heaters all around it, to heat it up. A quartz boat holding around 20 cleaned wafers is lowered into the hot zone, where a temperature of roughly 1000°C to 1200°C is maintained.

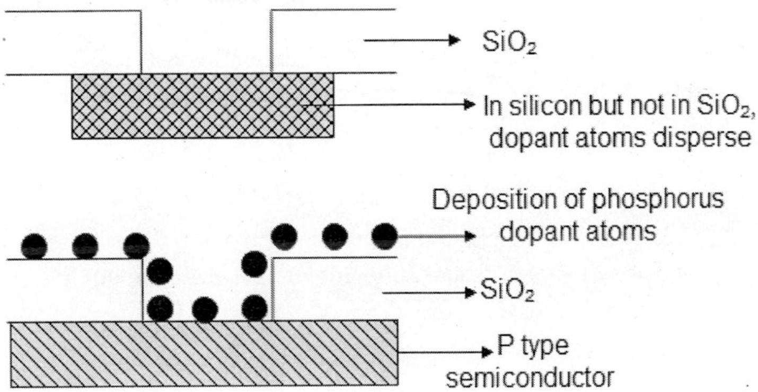


**Fig. 6.8:** Diffusion Process

Dopant addition or introduction of impurity atoms, alters the resistivity of silicon, in a semiconductor material. The dopants may be solid, liquid, or gaseous in nature. Pentavalent impurities or n-type dopants, such as antimony, phosphorous pentoxide, and arsine, are the preferred dopants. Gallium, indium, boron, and other p-type impurities are trivalent dopants. $B_2O_3$ (boron oxide), $BCL_3$ (boron chloride), and $P_2O_3$ (phosphorous oxide) and $POCL_3$ (phosphorous oxychloride) are examples of the impurities that spread. In diffusion process phosphorous oxychloride ($POCL_3$) is used as a safe liquid phosphorous source.

Next pollutants are transported to the high temperature zone using a carrier gas, such as $N_2$ or $O_2$. The temperature and diffusion time, affects the depth and width of diffusion. The usual direction of an impurity's diffusion is both laterally and vertically. Diffusion is the addition of impurity atoms from region of high concentration to low.

### 6.3.3.1 Diffusion Atomic Mechanisms

Boron (B), phosphorus (P), and arsenic (As) impurity atoms used as dopants. It occupies substitution locations, but dopant atoms might donate holes or free electrons to the lattice of silicon. Similar to the process where surplus carriers are formed in a device in a non-uniform manner, a carrier gradient is created when impurities diffuse into a solid. In each case, diffusion results from random motion, with particles moving in the direction of the gradient of decreasing concentration. Unless temperature is very high, the atoms impurity of the random motion in a solid is reduced. Impurities can infiltrate into the lattice by one of two physical methods. They are,

1. Diffusion of the substitutes
2. Diffusion in the interstitial zone

### 6.3.3.2 Diffusion of the Substitutes

At high temperatures, a semiconductor loses more atoms from its lattice locations, creating gaps that impurity atoms can fill as shown in Fig. 6.9 (a). This type of vacancy causes the impurities to disperse and, after cooling, these impurities occupy lattice positions in the crystal. It arises when the parent crystal's silicon atoms are replaced with doping impurities.

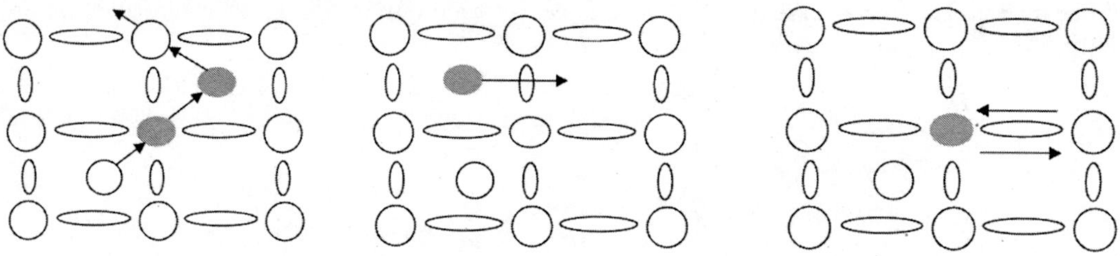

(a) (b) (c)

**Fig. 6.9:** Types of Physical Methods: (a) Diffusion of the Substitutes, (b) Diffusion in the Interstitial Zone, (c) Interstitial Movement

### 6.3.3.3 Diffusion in the Interstitial Zone

In this type of diffusion, there is no chance for the silicon atom to replace the impurity atom; instead, it diffuses into the lattice's interstitial spaces as shown in Fig. 6.9 (b). The most frequent pollutants that diffuse by this method are gold, copper, and nickel. In order to reduce carrier life time in silicon, which is crucial for boosting speed in digital ICs, gold is specifically employed. To further understand interstitial diffusion, think of a silicon unit cell with a diamond lattice and five interstitial spaces. There is an adequate space in each vacancy for an impurity atom. An impurity atom in one of these voids can move to a nearby vacancy as shown in Fig. 6.9 (c). The rate at which impurities diffuse into the semiconductor lattice is determined by the following factors:

- Diffusion mechanism
- Temperature
- Physical properties of impurities
- Lattice environment properties
- Impurity concentration gradient
- Geometry of parent semiconductor

## 6.3.4 Ion Implantation

Ion implantation is a method for introducing contaminants or for adding dopant impurities into a silicon wafer. It is done by, interject dopants in surface of wafer portions by blasting it using large energy dopant ions. The characteristics of the silicon wafer can be altered by the accelerated ions. The ion implantation setup is shown in Fig. 6.10. A beam (arc) of highly energetic dopant ions is used to scan silicon wafers inside a vacuum chamber (boron for the p-type and phosphorous for the n-type). The ions are then passed through a strong magnetic field after being accelerated in an electric field to energy of around 20kV to 250kV. The ions are applied with high energy and low temperature. undesirable impurities may be generated during arc discharge. To mitigate the presence of these impurities, a magnetic field can be employed to separate them from the desired dopant ions. The separation is based on the principle that the deflection of a particle in a magnetic field is influenced by its mass.

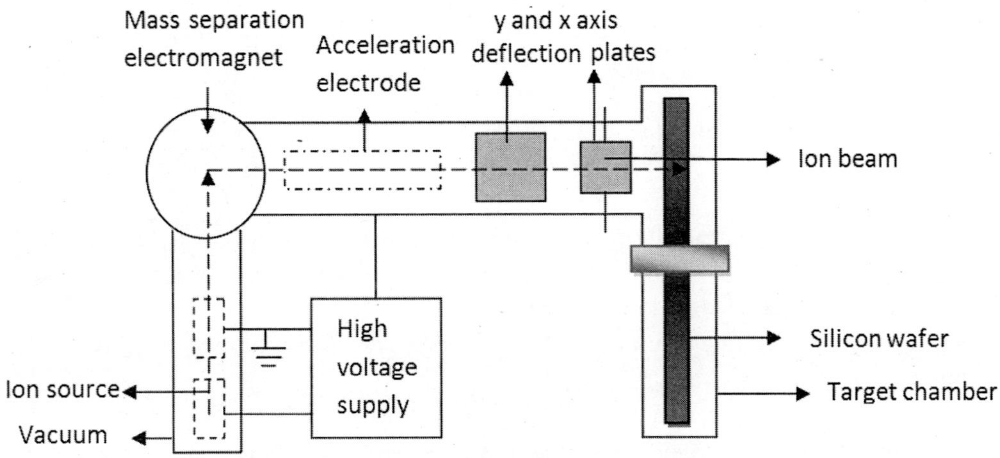


**Fig. 6.10:** Ion Implantation

The ions are further accelerated by the magnetic field's action until their energy reaches several hundred KV, at which point they are focused on and strike the target. The incident ion beam can be precisely measured as an electric current, making it possible to precisely adjust the doping level. Controlling the incident ion velocity makes it simple to control the depth of the dopant. It can only penetrate to a depth of a few inches. The ion implantation process in wafer level is depicted in Fig. 6.11.

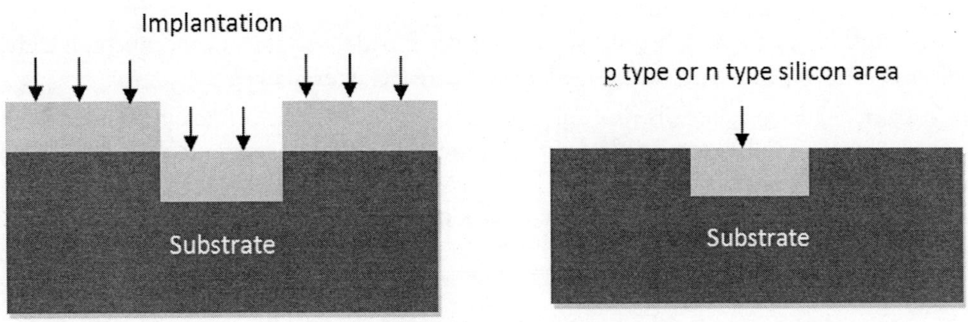


**Fig. 6.11:** Implantation Process

The dopant is guaranteed to be extremely pure. It is possible to precisely manage the doping homogeneity over the surface. The doping area may be precisely determined since the ions penetrate the solid as a focused beam with very minimal beam spread. The depth of the ions' penetration is determined by energy and accelerating field voltage. By changing the amount of ion concentration, the number of dopants can be managed. As a result of the controlled current, voltage, and energy, ion implantation is more accurate. The procedure is typically applied when the device requires a precise number of impurity atoms to function. The movement of contaminants is minimized because this is an all-low-temperature procedure. However, the extra energy may potentially harm the solid's crystal structure.

#### 6.3.4.1 Annealing

Due to high energy ions, the surface of the silicon wafer gets damaged in ion implantation process. In the surface region, this leads to the disarray of silicon atoms. The annealing process is used to get back the silicon wafer region, into a well-ordered crystalline structure. In the annealing process, the wafer is heated upto 1000°C for 30 minutes duration. There are two types of annealing techniques available. They are laser beam annealing and electron beam annealing. The wafer surface region is heated and re-crystallized in both annealing techniques.

#### 6.3.4.2 Advantages

Ion implantation is more expensive in terms of equipment cost and throughput. But it has the following advantages:

- Ion implanting takes place at low temperatures.
- During implantation, the beam current may be precisely measured, and a precise amount of impurity can be introduced.
- In most cases, the penetration depth will be between 0.1 and 1.0 mm.
- It is feasible to have extremely low dosage values while obtaining very big values of sheet resistance due to exact control of doping concentration.

### 6.3.5 Chemical Vapour Deposition (CVD)

One or more volatile precursors are introduced to a substrate, which reacts or degrades on its surface to produce the required thin film deposit. High-quality, high-performance solid materials are created via CVD, or chemical vapour deposition. In the semiconductor sector, it creates thin films. The procedure, also known as vacuum deposition, is carried out at a pressure lower than that of the atmosphere. On the surface of a substrate, the chemicals and vapours react to create solids. The CVD technique is used to deposit the protective layers on the substrate, such as silicon dioxide, polysilicon, and silicon nitride. For instance, $SiO_2$ is created when oxygen on the wafer surface reacts with silane gas.

Although the CVD technique falls short of thermally generated oxide in quality, it is sufficient to serve as an insulator on the substrate. The faster deposition rate at a lower temperature is its key benefit. Very low pressures, atmospheric pressures, sub-atomic pressures, and extremely high pressures can all be used for CVD. On a semiconductor device, contacts or plugs are also made using tungsten CVD. Chemical vapour deposition adds a variety of layers. Thermooxidation adds a $SiO_2$ layer to a silicon substrate.

Chemical Vapour Deposition (CVD) has grown in popularity and the primary method of deposition for a variety of materials. Atmospheric-pressure Chemical Vapor Deposition (APCVD), Low-pressure Chemical Vapor Deposition (LPCVD), Plasma-enhanced Chemical Vapor Deposition (PECVD) are the most prevalent CVD technologies. In APCVD, chemical vapour deposition (CVD) is done at atmospheric pressure. In LPCVD, CVD is done at sub-atmospheric pressures. This reduction in pressure reduces unnecessary gas phase reactions, thus increases film uniformity across the wafer. In PECVD, CVD is done by plasma to enhance the chemical reaction rates of the precursors. The deposition of plasma polymers is used for nano-particle surface functionalization, at lower temperature. The advantages of CVD are listed as follows:

- Low-pressure deposition methods
- Step coverage in uniform
- Accurate composition
- Structure control
- Process in low-temperature
- High deposition rates
- Processing cost is very law

The properties and applications of the three CVD techniques (APCVD, LPCVD and PECVD) are compared in Table 6.2. Furthermore, LPCVD does not require any carrier gases, which reduces particle pollution. The most significant disadvantage of LPCVD and APCVD is their high working temperatures, and PECVD is an effective way to address this issue.

CVD is the breakdown of gaseous compounds into stable solids using heat, UV, RF, plasma, or a combination of sources. Chemical vapor deposition is also known as Vapor Phase Epitaxy (VPE). Gaseous chemical processes are used to create an epitaxy layer on a heated surface. The change in free energy that occurs as a result of a chemical reaction is a plausible force of nature deposition.

Various chemistries including hydride, halide, and organometallic can be used in this VPE method. The silicon semiconductor industry is familiar with the halide and hydride systems, as epitaxial layers are commonly deposited via hydrogen reduction of chlorosilanes or pyrolytic decomposition of silane. The flow diagram of the processing steps of VPE growth method is shown in Fig. 6.12.

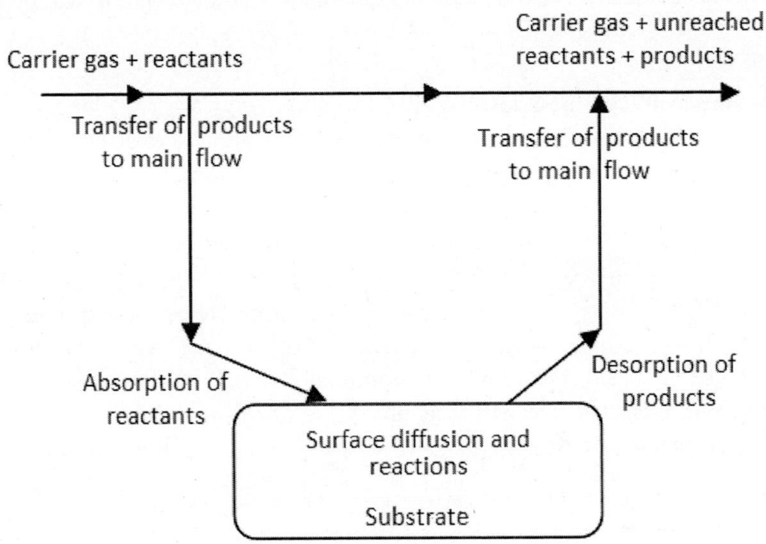

**Table 6.2:** Comparison between APCVD, LPCVD and PECVD Techniques

| S.No | Parameter | APCVD | LPCVD | PECVD |
|---|---|---|---|---|
| 1 | Reaction Type | Heterogeneous | Heterogeneous | Heterogeneous |
| 2 | Limiting Factor | Mass Transport Rate | Surface Reaction Rate | Surface Reaction Rate |
| 3 | Energy | Thermal | Thermal | RF Reaction Rate |
| 4 | Temperature Range | 550°C – 800°C | 550°C – 800°C | 200°C – 450°C |
| 5 | Pressure Range | 100–760 Torr | 0.25 – 1.0 Torr | 0.25 – 1.0 Torr |
| 6 | Chamber Walls | Hot / Cold | Hot / Cold | Cold |
| 7 | Step Coverage | Conformal | Conformal | Non-conformal to near-conformal |
| 8 | Range of Materials | Limited | Limited | Large |
| 9 | Film Purity | Lower | Excellent | Lower |
| 10 | Deposition Rate | Higher | Lower | Higher |

**Fig. 6.12:** VPE Growth Method

In Fig. 6.12, the reactants are first introduced in VPE growth method, and transferred into the substrate region. Then the absorption of reactants takes place on the substrate surface. On the substrate surface, surface reactions such as surface diffusion, site accommodation, chemical reactions, and layer deposition occur. On the surface of the substrate, desorption of residual reactants and by-products takes place. At last, from the substrate surface transfer and removal of residual reactants and by-products done.

## 6.3.6 Sputtering

Sputter is one of the Physical Vapour Deposition (PVD) method, which is used for thin film deposition. This is done by the phenomenon called sputtering. Sputtering is a highly controlled process that can be used to deposit films of metals, insulators, semiconductors, and alloys in microelectronic production. It has better step coverage than evaporation, causes significantly less radiation damage than electron beam evaporation, and produces layers of composite materials and alloys far better.

Sputtering is the process of surface atoms being ejected off an electrode surface by transferring momentum from bombardment ions to the electrode surface atoms. The simple sputtering system setup is shown in Fig. 6.13. The electrode material is placed on the substrate. The chamber is usually supplied with an inert gas. The chamber's gas pressure is maintained at around 0.1 torr. Sputtering can be used to deposit a wide range of materials because of its physical nature. Simple DC sputtering is commonly preferred for elemental metals due to its high sputter rates. RF plasma must be utilised for installing insulating materials like $SiO_2$.

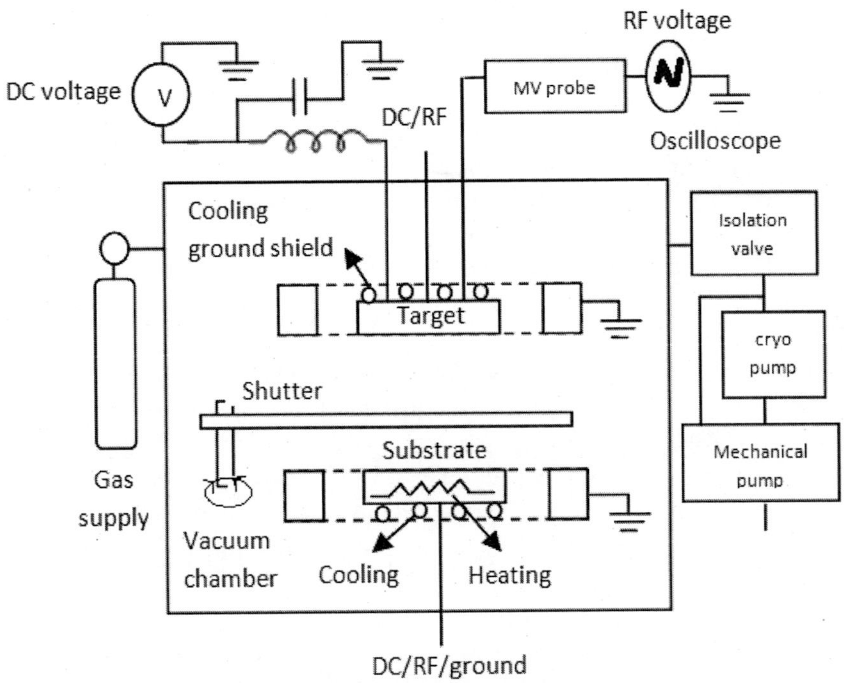


**Fig. 6.13:** Simple Sputtering System

## 6.3.7 Photolithography

The method of defining patterns by smearing a thin, even coating of viscous liquid (photo-resist) onto the surface of the wafer is called Lithography. Baking hardens the photo-resist, which is then removed with precision by shining light through a reticle that contains the mask data. Lithographic Technologies are employed in the processing of semiconductors such as ion lithography, x-ray lithography, photolithography, and electron lithography. Photolithography is a method for creating microscopic-scale patterns for devices and circuits on silicon wafers. Device dimensions or line widths as small as 25 nm can be obtained using the photolithographic technique, which exposes objects to UV radiation. Two processes are involved in photolithography. They are,

- Making a mask for photolithography
- Photo etching

### *6.3.7.1 Photolithography Mask*

An area-specific thin layer of an opaque material has been deposited on a flat piece of transparent glass to create the desired pattern. Tools are utilised to create the pattern for the mask, which is made using lithography.

### *6.3.7.2 Photo Resist*

A photo resist is a radiation-sensitive chemical that, when exposed to radiation, generates a polymer coating. When exposed to UV rays, it significantly rises or diminishes. The wafer is thoroughly cleaned to encourage resistance towards wetting and adherence. To eliminate solvents, encourage 10–20 minutes at 90°C (190°F) in temperature. The resist is visible through the pattern mask is aligned with the wafer. Photo resists are classified as negative or positive depending on the solubility changes that occur. In negative photo resists, solidified materials are less soluble in developer solutions. So negative masks are created by brightness and have a negative pattern. In the case of positive photo resists, brightness makes the exposed region more soluble, allowing it to be removed more easily during the developing process. As a result, the patterns created in the positive resist on the wafer are identical to those on the mask, and the patterns etched are the opposite of mask. The change in solubility is caused by a series of chemical processes that take place as the substance dissolves.

### 6.3.7.3 Etching

Any remaining layer of resist on surface is stripped using either liquid chemicals or plasma etching. Etching is the process of removing a target material using an aqueous solution, typically an acid. The etchant is chosen to chemically attack the specific material to be removed rather than the protective layer. The overall photo etching steps are shown in Fig. 6.14. They are,

i. First, photo resist film of thickness 5000 – 10000 Å is applied on the silicon wafer.
ii. The negative photo mask is created on photo resist film to create the required pattern and UV rays are applied. Under the mask transparent, region photo resist becomes polymerized.
iii. Now the mask is removed using chemicals, wafer is developed. It gives the pattern as in Fig. 6.14 (c), by dissolving on the photo resist unpolymerized regions.
iv. The areas not protected by photo resist are removed by etching solution of hydrofluoric acid. $SiO_2$ region is removed in particular portion by using photolithography which is shown in Fig. 6.14(d).

After contaminants have diffused, the photo resist is removed using a combination of hot sulfuric acid and mechanical abrasion. Wet and dry etchings are the two types of chemical etchings. In wet etching, the immersed material is dissolved in a chemical solution. In dry etching, using reactive ions or vapour phase etchant, the material is dissolved or sputtered.

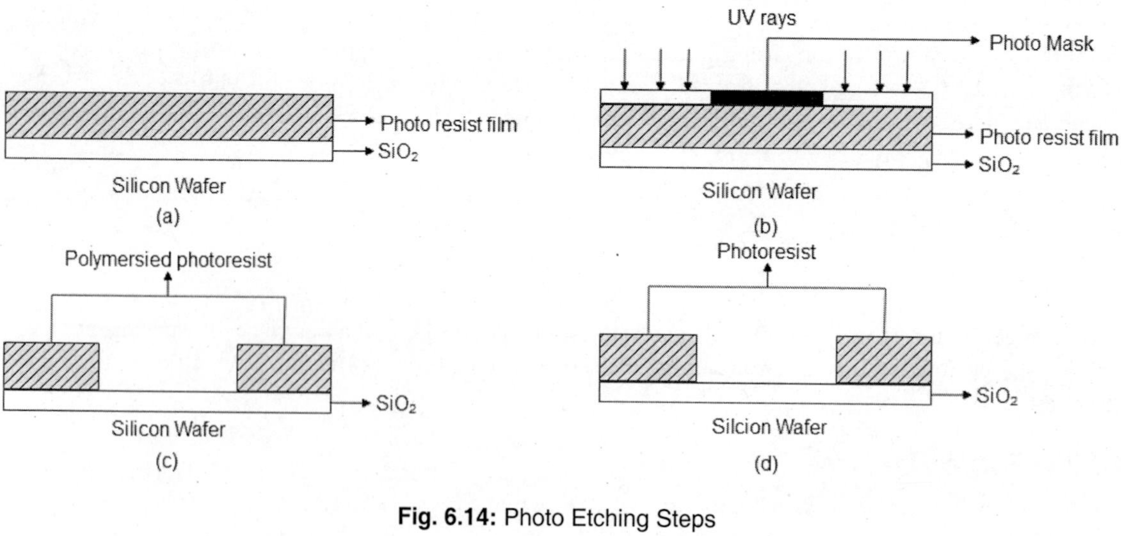


**Fig. 6.14:** Photo Etching Steps

### 6.3.7.4 Wet Etching

During wet etching, the wafers are immersed in a chemical solution at a particular temperature. All directions of the material to be etched are equally eliminated throughout this procedure, which causes some material to be etched from areas where it is intended to be left.

The wet etching method is the oldest etching method. It removes the thin film that isn't protected by photo resist by using a chemical interaction between the thin film and solvent. There are two types of wet etching. They are,

- Anisotropic etching
- Isotropic etching

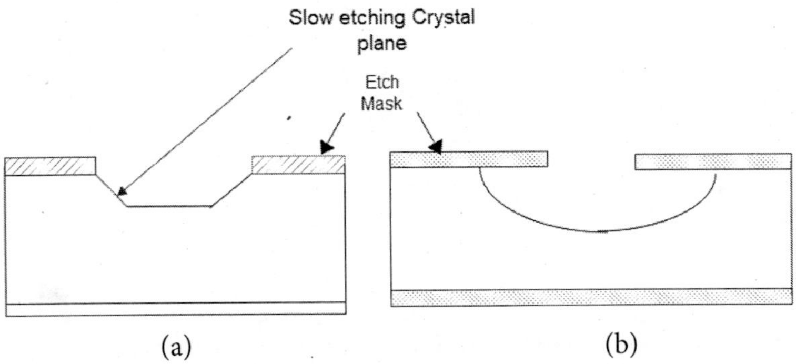


**Fig. 6.15:** Wet Etching: (a) Anisotropic Etch, (b) Isotropic Etch

Anisotropic etching is direction-dependent, whereas it eliminates material along a certain direction is shown in Fig 6.15 (a). The etching rate in isotropic etching is constant in all directions, so a material's attributes are unaffected by the direction as shown in Fig. 6.15 (b). Silicon dioxide, single-crystal silicon, silicon nitride, and metal are all widely etched using the wet etching method. When compared to the dry etching method, the wet etching process has a higher throughput and is normally isotropic, though it can also be anisotropic. As a result, it is unsuitable for establishing line features in contemporary deep submicron techniques. As shown in Fig. 6.16, two steps are involved in wet etching. They are,

1. Reactant movement towards the responding surface (Ex: diffusion)
2. Surface and surface-level chemical reactions

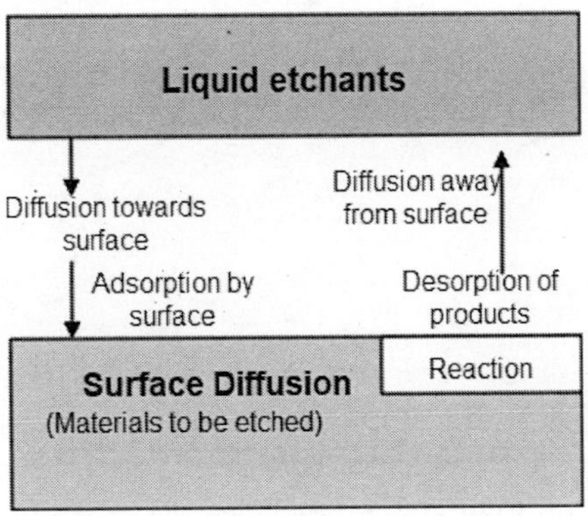


**Fig. 6.16:** Process of Wet Etching

### 6.3.7.5 Dry Etching

Wafers are submerged in gaseous plasma generated by RF electric field being applied to a gas. Ex: argon, during dry (or plasma) etching. The electrodes absorb the field's kinetic energy collide with the gas molecules, and transmit energy to them, producing ions and electrons as a result. Avalanche process spreads the entire gas, the newly created electrons clash with other gas molecules to form plasma. On an electrode, a wafer that needs to be etched is placed before being exposed to gas ions being fired at its surface. Ions are used to remove atoms from the surface of a material. The momentum transfer from the ions to the atoms at the etching surface plays a crucial role in this removal process. The ionized gas is produced by placing the right gas mixture within a vacuum chamber.

Since the feature size reached 3 nm in the late 1980s, the dry etching (also known as plasma etching) technique has steadily supplanted the wet etching process for all patterned etching operations. The dry etching procedure has become the primary etch approach in semiconductor fabrication because of its good anisotropic profile and ability to generate highly reactive chemical species. It can remove the material solely through chemical reactions with chemically reactive gases or plasma, purely physical processes like sputtering and ion beam induced etching, or a combination of chemical reaction and physical bombardment. A parallel-plate reactive ion etching system is shown in Fig. 6.17. Dry etching is a typical method for etching dielectric, single-crystal silicon, poly silicon, and metal, as well as stripping photo resist.

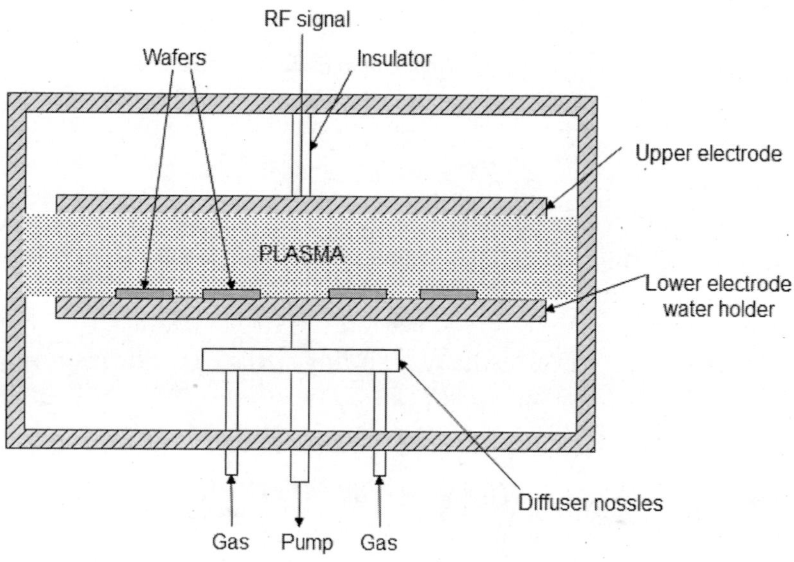


**Fig. 6.17:** Parallel-Plate Reactive Ion Etching System

### 6.3.7.6 Sputter Etching Process

Ions are used in another plasma etching procedure called the sputter etching process. Ions driven by a high electric field, assault the atoms on the wafer surface in this etching process, physically dislodging them. More physical etching components, such as sputtered surface materials, come from this bombardment. The substance evaporates when the high-energy particles knock away the atoms from the substrate surface. Because of no chemical reaction, only the item that has been unmasked will be removed.

It has weak selectivity but a high direction, i.e., an anisotropic profile, because it is primarily a physical reaction. Sputtering-etching system, that uses noble gas ions with relatively high energies (>500 eV). Argon ions are accelerated by the supplied electric field and bombard the target surface. Atoms near the surface are sputtered off surface due to momentum transfer. From 0.01 to 0.1 torr is a common working pressure for sputter etching.

### 6.3.7.7 Optical Lithography

Photolithography is also known as optical lithography, if it uses light to create thin films. On the substrate, such as silicon wafers, it is utilised to shape thin sheets. Throughout subsequent production procedures, it shields particular locations (deposition, ion implantation, etc.). UV light, X-rays, and severe UV that radiate at various frequencies are the types of light utilized to introduce mask patterns on the silicon wafer as shown in Fig. 6.18. Ultra-violet light is the most common form of light used to create films. Photo resists, a substance that is light-sensitive but becomes rigid when exposed to UV radiation.

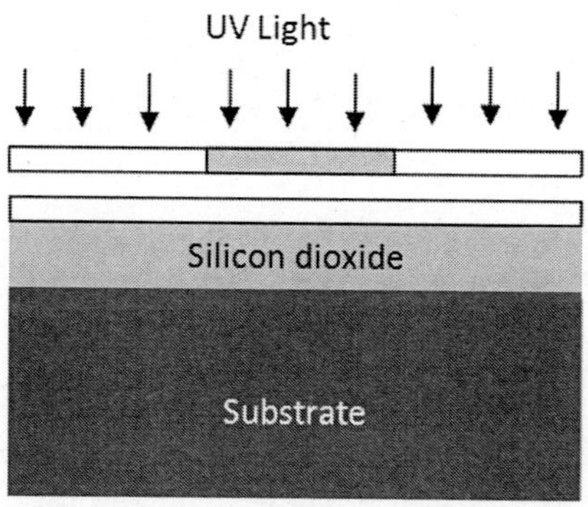


**Fig. 6.18:** UV Light Impinging on the Substrate

A photo resist is first applied to the silicon wafer or substrate. The silicon wafer is covered with a photolithographically applied mask pattern on the photo resist layer. The wafer's exposed regions soften and become removable. On the wafer, it makes a pattern. The size of the features imprinted on the photo resist depends on the wavelength of each type of light. Because the exposed region is patterned with the aid of light, photolithography and photography share some similarities.

Through the lens, the incident light may be direct or projected. A mask is used to expose the wafer to UV light. It transfers the geometric pattern from the mask to the photo resist-coated wafer's surface. The photo resist either breaks or hardens when the UV light strikes the exposed area on the substrate. Using the particular solvent, the coating's softened areas are further removed. Microprocessors and solid-state devices are frequently made via optical lithography. It has three types. They are,

i. Printing contact
ii. Proximity printing
iii. Projection printing

i. **Printing Contact**

Printing contact is the most commonly used method. It is done by pushing the mask against the resist coating during exposure as shown in Fig. 6.19 (a). Using this, high resolution of the pattern, onto the wafer surface is produced. However, the mask eventually degrades when it comes into contact with the wafers. So, this method gives low yield.

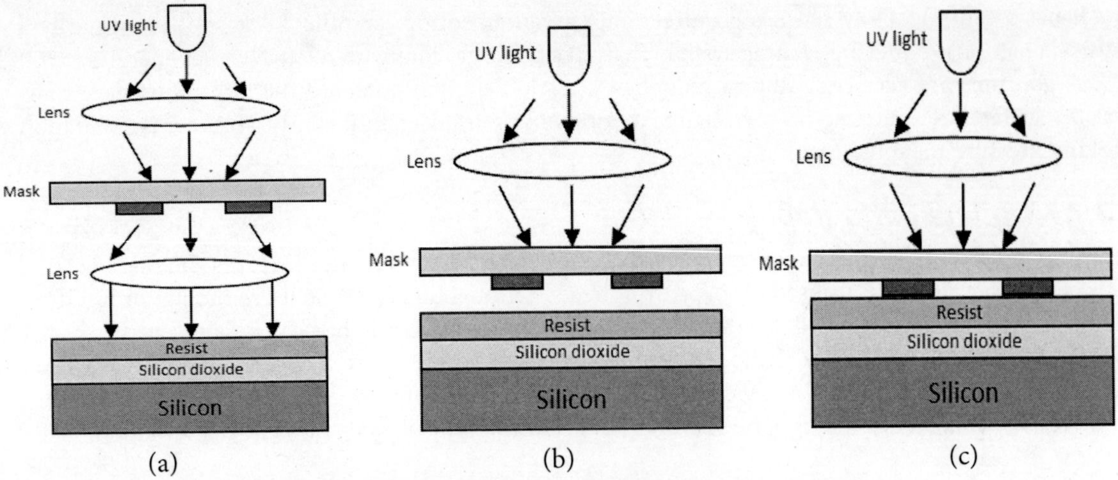


**Fig. 6.19:** Optical Lithography Types: (a) Printing Contact, (b) Proximity Printing, (c) Printing for Projection

ii. **Proximity Printing**

In proximity printing, there is no contact between mask and wafer as shown in Fig. 6.19 (b). A distance of 10 to 25 μm separates the mask from the resist coating eliminating the need to wear a mask. But it leads to higher yield and lower resolution due to diffraction.

iii. **Projection Printing**

In modern optical lithography, Projection Printing is mostly used. The image through the mask is projected onto the wafer by an improved quality lens system as shown in Fig. 6.19 (c). It is a preferred method because there is no contact (requiring no mask wear) and it produces good resolution.

## 6.3.8 Metallization

Metallization is a procedure that connects components together. After all of the semiconductor fabrication stages for a device or an integrated circuit have been finished, metallic connectors for the integrated circuit, as well as external connections to both the device and the IC, are required. The process of applying a metal layer to a metallic or non-metallic surface is known as "metallization". Silver, zinc, or aluminium can be used as the coating in metallization process. Aluminium is the metal, mostly used in CMOS production to protect the surface from environmental elements like water, air, and dust.

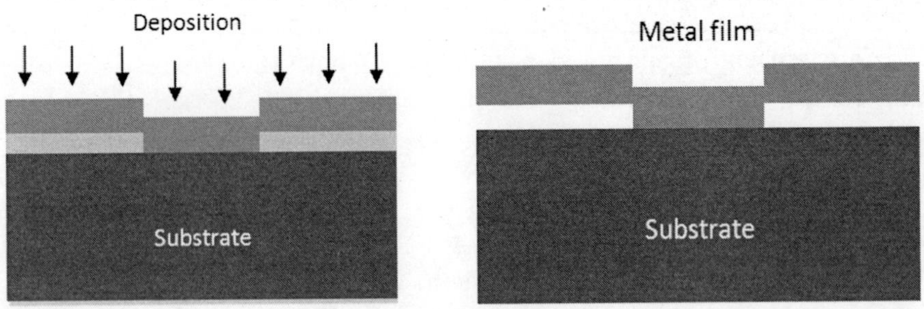


**Fig. 6.20:** Metallization Film

Many components that make up an integrated circuit are also connected using metallization. The parts can be relays, transistors, capacitors, resistors and more. A chamber is used in the metallization process to apply the metal layer. In general, the substrate's whole surface is coated by the aluminium after it is placed inside the chamber by the vapour evaporation method. After this, the metal layer (metal film) is placed on the silicon wafer's surface as shown in Fig. 6.20.

Depending on the needs, the metal layer's thickness can change. Metallization creates the necessary extremely precise patterns of conductive material on wafer surface using photolithography. The unnecessary area for connection is then etched by using etchants, as shown in Fig. 6.21.

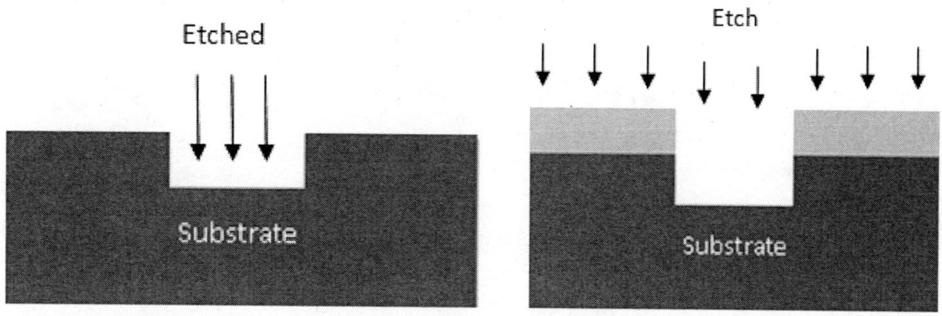


**Fig. 6.21:** Metallization Etch

## 6.3.9 Packaging

The final step in the manufacture of an IC is packaging. An element of a device is an IC. The ICs are put together with other components during packaging. A device is created by combining the IC's external connections. The packaging of a product serves as a link between its manufacture and final application. It prepares a product for its intended usage. To protect it from air dust, the box is eventually sealed with epoxy or plastic. An automatic probing station is used to conduct the tests. Microwave and radio-frequency testing are both available on this efficient testing device. The effective circuits are sent for headers or packaging. Dependability is the major design issue in IC packaging.

There are different IC packages and many classification schemes available. Depends on their mounting style, IC packaging can be categorized into two types. They are,

- Through-hole Technology
- Surface Mount Technology (SMT)

In a Through-hole package, for connecting the components, through-hole technology drills holes through the Printed Circuit Board (PCB). Leads on the component are mechanically and electrically connected to the PCB by being soldered to pads on the PCB. There are different types available in through-hole IC package as shown in Fig. 6.22. They are,

- Single In-line Pin Packages (SIP or SIPP)
- Dual In-line Packages (DIP)
- Zig-Zag in-Line Packages (ZIP)

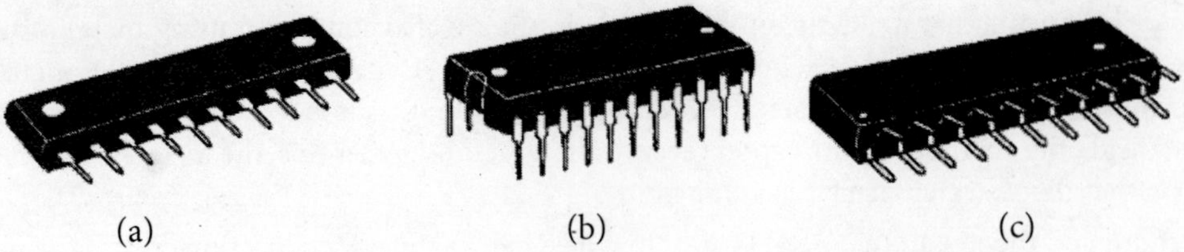


**Fig 6.22:** Through-hole Package: (a) SIP, (b) DIP, (c) ZIP

The direct mounting or placement of the electronic components on the printed circuit board's surface is known as the surface mount packaging technique. There are different types of surface mount packaging available as shown in Fig. 6.23. They are,

- Small Outline Integrated Circuit (SOIC)
- Small Outline Package (SOP)
- Quad Flat Pack (QFP)
- Plastic Leaded Chip Carrier (PLCC)
- Ball Grid Array (BGA)

**Fig. 6.22:** Surface Mount package: (a) SOIC, (b) SOP, (c) QFP, (d) PLCC, (e) BGA

For IC packaging, plastics or ceramics are the materials that often get used, due to its better electrical conductivity. Dual In-line plastic Package is the most commonly used IC packaging technology.

## 6.4 TWIN TUB METHOD

The twin tub method is one type of CMOS fabrication process. In CMOS fabrication, implantation of two distinct tubs with lightly doped silicon is known as the twin tub process. The twin-tub, as its name suggests, is a combination of p-well and n-well processes created on the same substrate. It is also called as Duel-well process.

The CMOS inverter by twin tub fabrication process is shown in Fig. 6.24. This creates an n-type substrate with high resistivity that contains both n-well and p-well regions. This distinct transistor and its configuration aid in the optimisation of the n-type and p-type devices, as well as other factors like body effect and threshold voltage. The wafer is made up by two layers. The main substrate layer is an n-type layer, while the top layer is epitaxial.

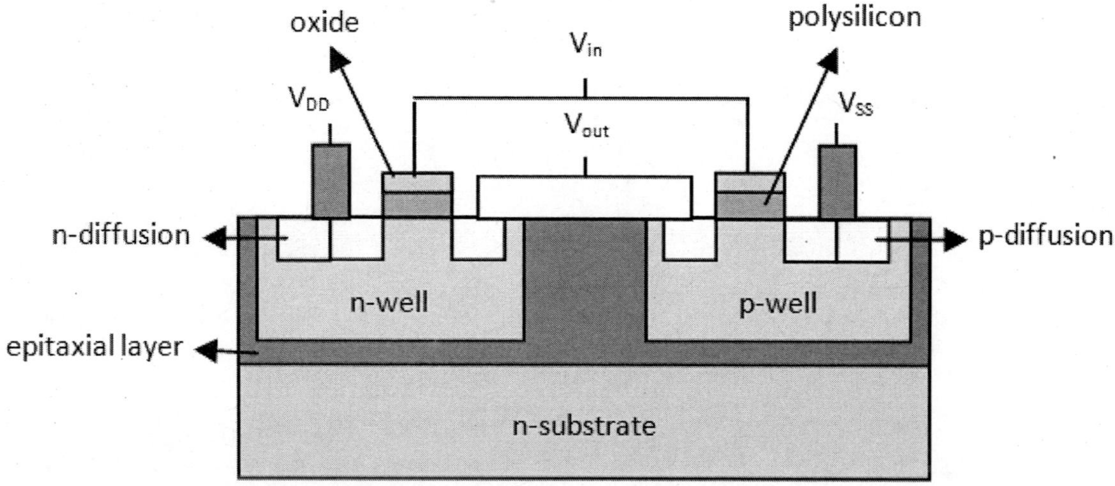


**Fig. 6.23:** Fabrication of an Inverter Utilizing Twin Tubs

There are various processes in the twin tub method used in CMOS fabrication. They are,

a. Lightly doped $n^+$ or $p^+$ substrate with a "epitaxial" or "epi" layer to protect the "latch up" is the starting material.

b. Epitaxy:
   1. Create controlled-thickness layers of high-purity silicon
   2. Grown silicon, needs to have accurately calculated dopant concentrations
   3. The dopant and its concentration in silicon determine the electrical characteristics

c. The following procedures need to be done.
   1. Tub formation
   2. Thin-Oxide construction
   3. Source and drain implantations
   4. Contact cut formation
   5. Metallization

CMOS fabrication utilising the twin tub process method by following the above procedures. The key benefit of this method is the ability to individually optimise the transconductance, body effect parameter, and threshold voltage.

## 6.5 EXAMPLE FOR CMOS IC FABRICATION STEPS

The circuit diagram of the CMOS inverter is shown in Fig. 6.25. In this, the gate terminals of both PMOS and NMOS is connected. The drain terminals of both PMOS and NMOS are connected together.

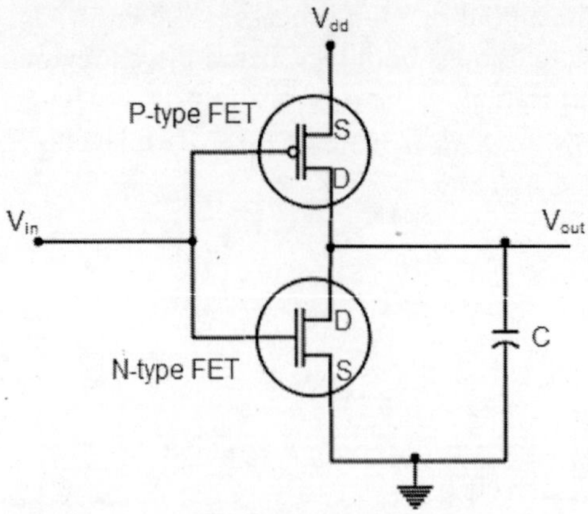


Scan to know more

**Fig. 6.24:** Schematic of CMOS Inverter Circuit

The overall step by step procedure for fabrication of CMOS Inverter using N-Tub fabrication is explained as follows:

| | | |
|---|---|---|
| | Create a blank wafer of p-type substrate as a base for fabrication. |  |
| 2. | Oxidation: $SiO_2$ oxide layer is created by using an oxidation process. $SiO_2$ production at 900°C–1200°C with $H_2O$ or $O_2$ on top of Si wafer at the oxidation furnace. | 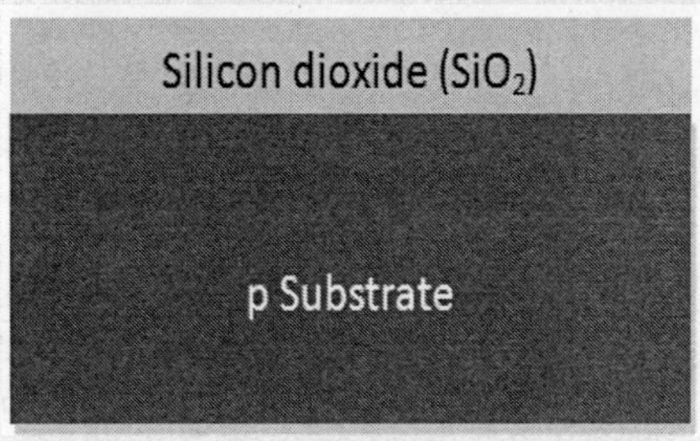 |
| 3. | Growing of photoresist: To allow selective etching in photolithography process, photoresist which is a light-sensitive organic polymer is covered over the $SiO_2$. It softens when exposed to light. | 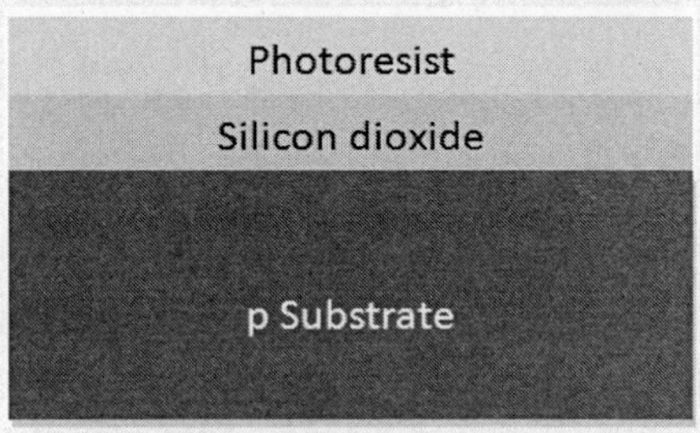 |

| | | |
|---|---|---|
| 4. | Masking: Create a mask to form N-well in p-substrate and expose UV rays. The photoresist under the mask gets polymerized. | 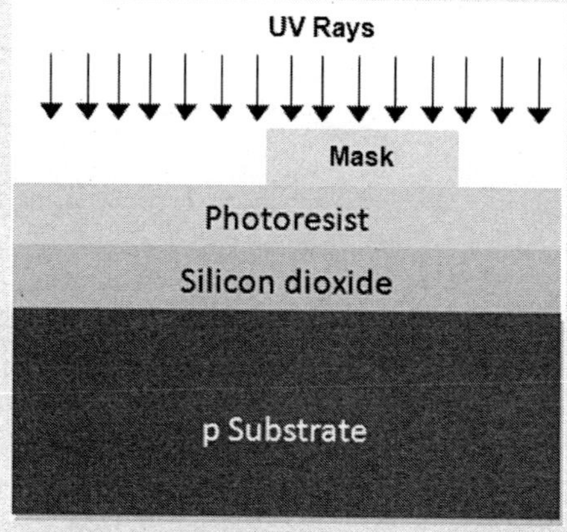 |
| 5. | Unexposed photoresist removal: Now the mask is removed and then the unexposed photoresist is dissolved. | 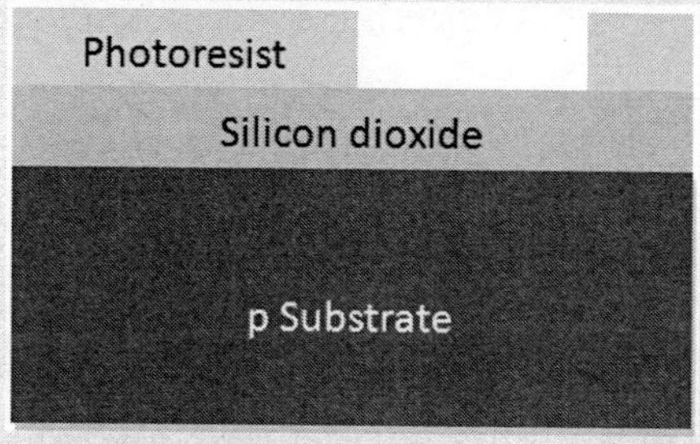 |
| 6. | Etching: By using Hydrofluoric acid (HF) etching solution, $SiO_2$ layer is removed in the selected area where dopants are to be diffused. Only oxide exposed to HF assaults when used to etch it. | 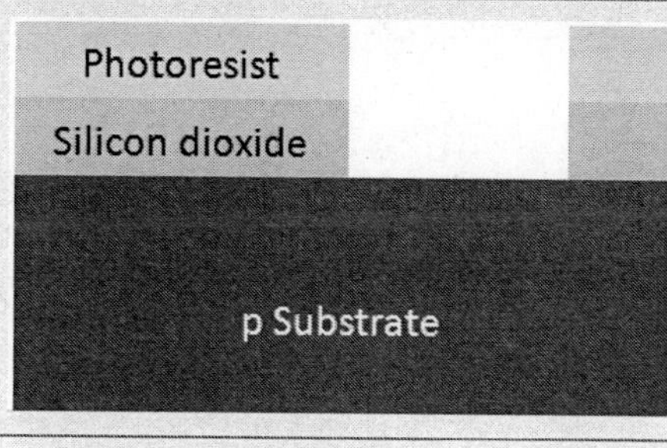 |
| 7. | Photoresist layer removal: Photoresist Strip, remove any leftover photoresist. Use the piranha etch acid mixture. This is done to prevent the photoresist from melting in the upcoming phase. | 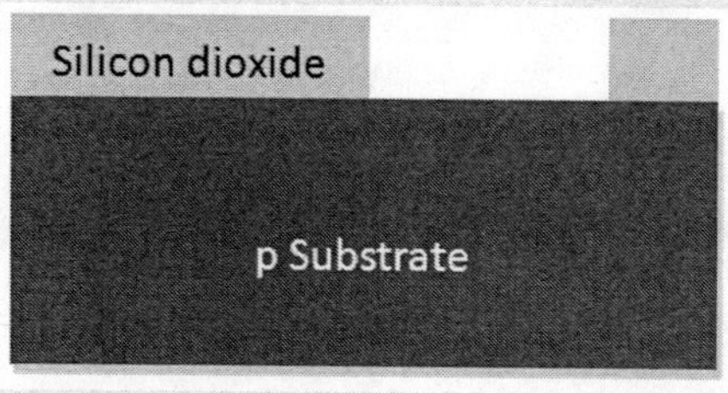 |
| 8. | N-well formation: To get n-well, n-type impurities are diffused into the p-type substrate. For that, either ion implantation (or) diffusion process can be used. | 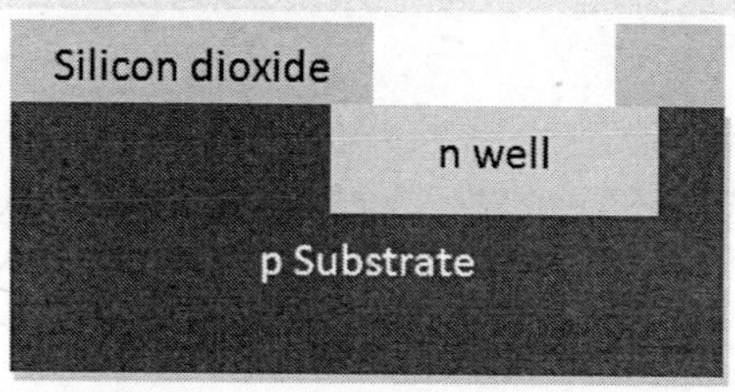 |

| 9. | $SiO_2$ removal: Strip the $SiO_2$ oxide layer using Hydrofluoric acid (HF). After this, NMOS can be formed over p-substrate and PMOS can be formed over n-well. | 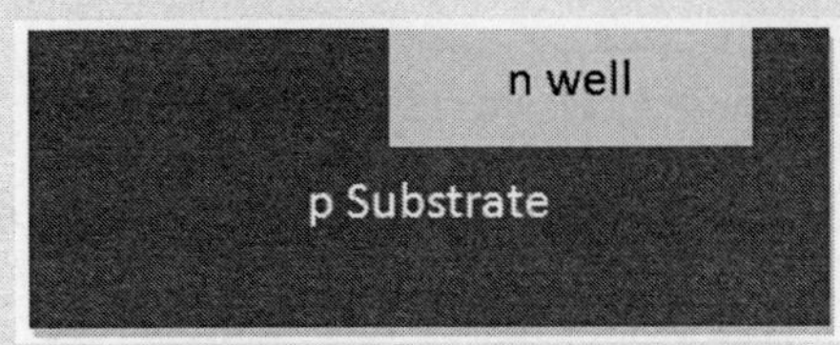<br> |
|---|---|---|
| 10. | Polysilicon deposition: By using chemical vapour deposition (CVD), 6-7 atomic layers of polysilicon which have a very thin layer of gate oxide ($SiO_2$) is deposited over substrate. The self-aligned gate process is preferred to prevent the misalignment of the CMOS transistor gate. This avoids the unwanted capacitance. This is done by formation of gate regions before forming source and drain region using ion implantation. | 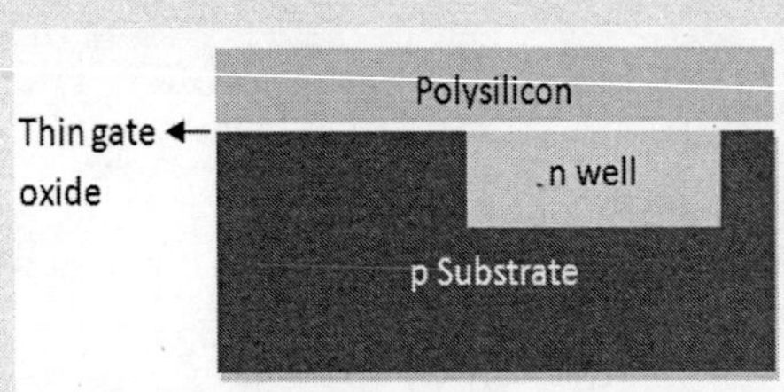<br> |
| 11. | Gate region formation: Two regions are required for the formation of gate such as PMOS and NMOS transistors. So, the remaining portion of polysilicon is stripped off by using the lithography method. | 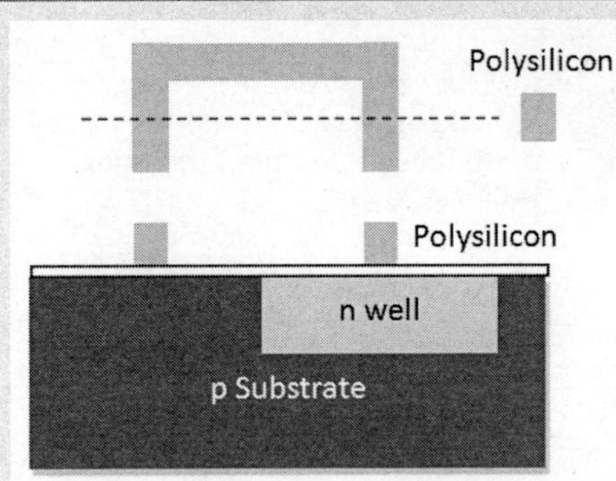<br> |
| 12. | Oxidation process: An oxide layer is deposited over the entire region. It acts as a shield for further diffusion and metallization processes. | 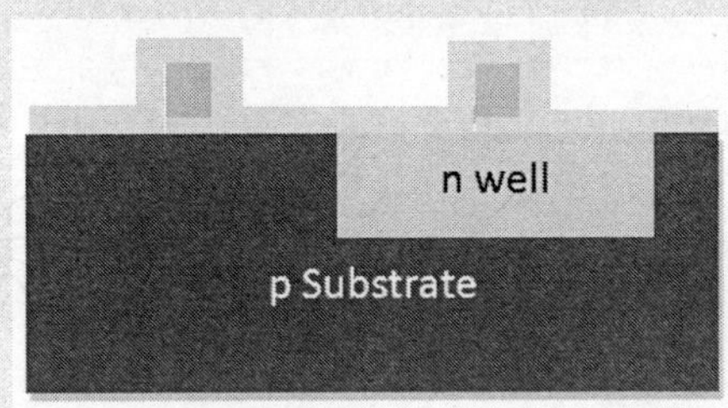<br> |
| 13. | Masking and Diffusion: By using the lithography masking process, small gaps are made for n+ diffusion. Since polysilicon doesn't melt during later processing, it is preferable to metal for self-aligned gates. | 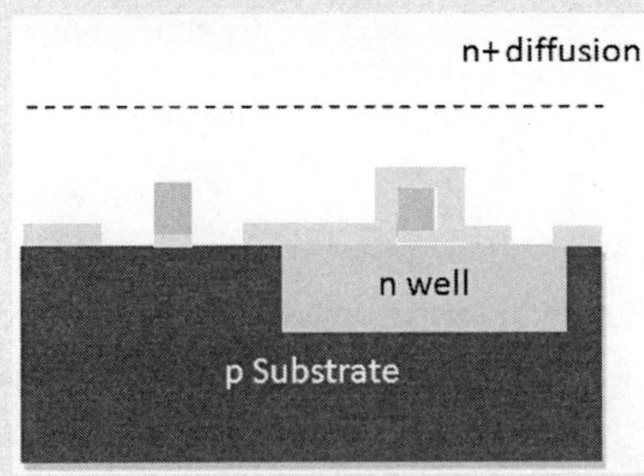<br> |

| | | |
|---|---|---|
| 14. | N+ diffusion: Using the diffusion process, three n+ regions are created. Now the NMOS transistor is formed. | 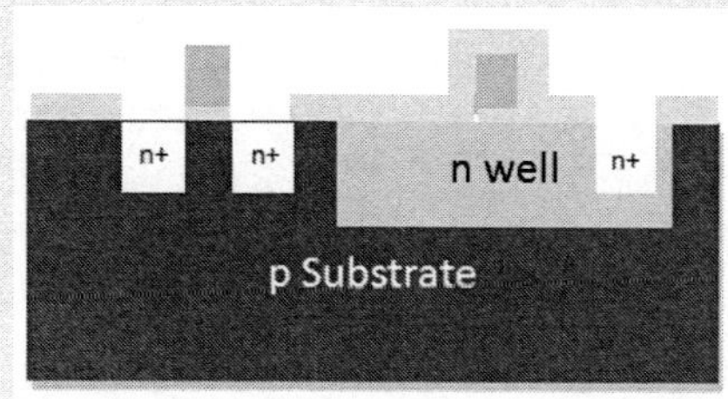 |
| 15. | Oxide removal: Strip off oxide to finish the patterning stage for n+ diffusion. | 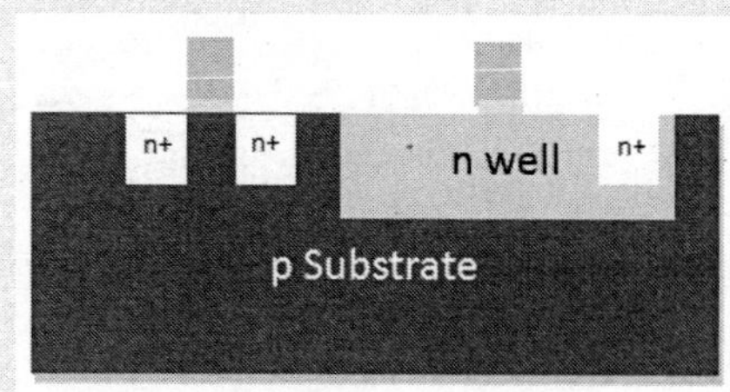 |
| 16. | P+ diffusion: Follow the steps 12, 13, 14 and 15 and do oxidation, masking and diffusion process. P+ impurities are diffused in three regions. Now the PMOS transistor is formed. | 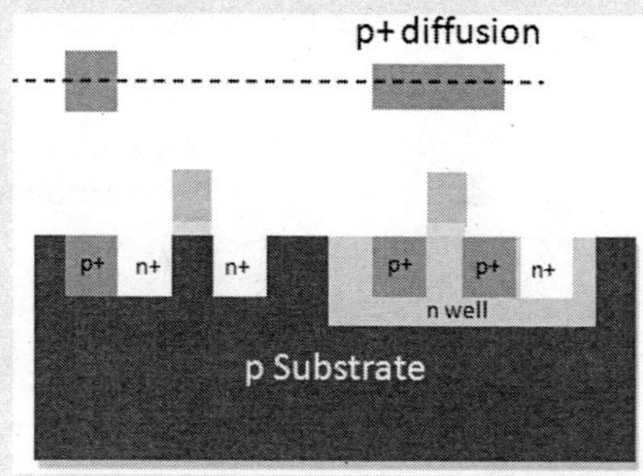 |
| 17. | Thick field oxide formation: Before attaching the components, cover the chip with a thick layer of field oxide and etch the oxide where contact cuts are necessary. This thick oxide layer acts as a protective layer where terminals are not required. | 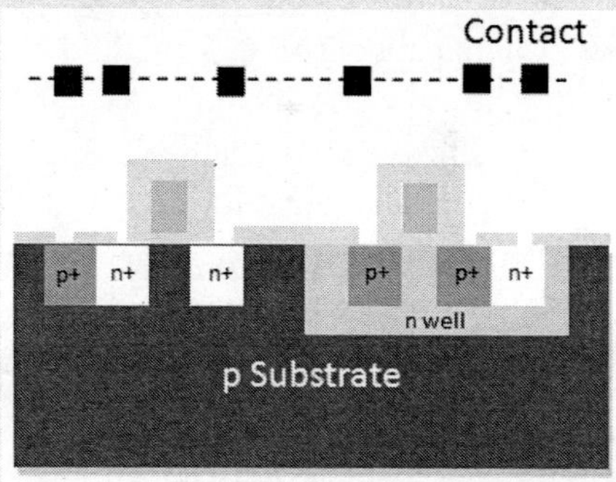 |
| 18. | Metallization: To create interconnections, sputtering aluminium over the entire wafer during metallization process. Remove superfluous metal using the pattern, leaving the wires. | 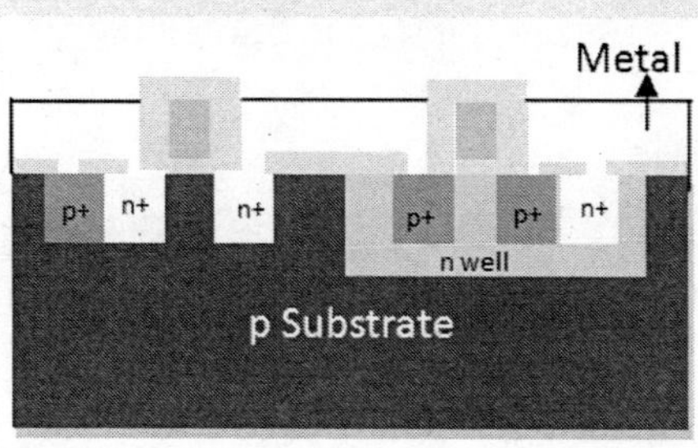 |

## UNIT SUMMARY

- IC is a collection of electronic devices, fabricated on a small chip of semiconductor material.
- ICs are small size, low cost and have lower power consumption.
- Silicon (Si) and Germanium (Ge) can be used for IC fabrication. But Silicon is most preferable due to its easy availability and advantages.
- Moore's law states that the number of transistors per chip grows exponentially or doubles in every 18 months.
- ICs can be classified based on fabrication method / technique, integration scale / level and application.
- Monolithic IC fabrication consists of different steps.
- Wafer preparation: Selecting and preparing the Si wafer for IC fabrication.
- Oxidation: Silicon dioxide is created during the oxidation process by the conversion of silicon layers on top of the wafer to oxygen (dry oxidation) or $H_2O$ (wet oxidation) molecules.
- Diffusion: Addition of impurity from high concentration to low concentration region.
- Ion implantation: The most common method for adding dopant impurities to semiconductors. Through the use of an electrical field, the ionised particles are accelerated and directed at the semiconductor wafer.
- Deposition: On the wafer, films made of different materials are applied. Physical vapour deposition (PVD) and chemical vapour deposition are the two main types of methods used for this (CVD).
- Lithography: The method of defining patterns by smearing a thin, even coating of viscous liquid (photo-resist) onto the surface of the wafer. Baking hardens the photo-resist, which is then removed with precision by shining light through a reticle that contains the mask data.
- Etching: Selectively cleaning the wafer's surface of undesirable material. Etching agents are used to impart the photo-resist pattern on the wafer.
- Chemical Polishing: A method of planarization that involves covering the wafer surface with a chemical slurry including etchant chemicals.
- Metallization: By using metal, connecting the components together is called as metallization. Here metal layer is formed on the top of the wafer.
- Packaging: ICs are put together with other components during packaging.

## EXERCISES

### Multiple Choice Questions

6.1 Advantage of integrated circuit is ________________

a. High reliability  b. Low reliability
c. High power  d. High cost

6.2 Silicon processing involves turning _________ into very pure silicon

a. Iron  b. Sand
c. Steel  d. Wood

6.3 A __________ wafer is a thin crystal semiconductor used for IC fabrication

a. Al  b. As
c. Si  d. Ge

6.4 The process of introducing oxygen is called __________

a. Deposition  b. Lithography
c. Diffusion  d. Oxidation

6.5 __________ is the addition of impurity atoms from a high concentration region to a low concentration region

a. Diffusion  b. Lithography
c. Deposition  d. Oxidation

6.6 Ions are accelerated from the element to the solid target during the __________

a. Diffusion  b. Ion implantation
c. Deposition  d. Oxidation

6.7 The __________ technique is used to deposit the protective layers on the substrate, such as silicon dioxide, polysilicon, and silicon nitride

a. OVD  b. IVD
c. CVD  d. PVD

6.8 The method of defining patterns by smearing a thin, even coating of viscous liquid (photo-resist) onto the surface of the wafer is called __________

a. Deposition  b. Oxidation
c. Diffusion  d. Lithography

6.9 A __________ is a radiation-sensitive chemical that, when exposed to radiation, generates a polymer coating

a. Photoresist  b. CVD
c. Diffusion  d. Lithography

6.10 __________ is the process of removing specific areas of a silicon dioxide, metal, or semiconductor

a. Photoresist  b. Etching
c. Diffusion  d. CVD

6.11 The process of applying a metal layer to a metallic or non-metallic surface is known as __________

a. CVD  b. Etching
c. Metallization  d. Oxidation

6.12 Photolithography process includes__________

a. Oxidation  b. Photo etching
c. Photographic  d. Both b & c

6.13 A __________ wafer is a thin crystal semiconductor used for IC fabrication

a. Lower than final chip dimension  b. Higher than final chip dimension
c. Both a & b  d. None

6.14 Which etching process is adoptable type for making photoresist?

a. Plasma etching  b. Wet etching
c. Both a & b  d. None

6.15 Which IC type is most commonly used?

a. Hybrid  b. Monolithic
c. Thin film  d. Thick film

6.16 Monolithic IC includes__________ components

a. Passive  b. Active
c. Both a & b  d. None

6.17 Oxidation is mainly used for ____________

a. Doping  b. Isolation
c. Interconnection  d. None

6.18 The performance of monolithic IC depends on ____________

a. Packaging  b. Substrate
c. Interconnection  d. Not substrate

6.19 We can use Metallization for __________

a. Packaging  b. Protection
c. Interconnection  d. None

6.20 We can use Optical masking for ________

a. Cleaning  b. Protection
c. Transfer of Pattern  d. None

| Answer of Multiple Choice Questions | | | | |
|---|---|---|---|---|
| 6.1 **a** | 6.2 **b** | 6.3 **c** | 6.4 **a** | 6.5 **b** |
| 6.6 **c** | 6.7 **a** | 6.8 **b** | 6.9 **c** | 6.10 **a** |
| 6.11 **b** | 6.12 **d** | 6.13 **b** | 6.14 **a** | 6.15 **b** |
| 6.16 **c** | 6.17 **b** | 6.18 **b** | 6.19 **c** | 6.20 **c** |

## SHORT AND LONG ANSWER TYPE QUESTIONS

### Category I

6.1 List out the step of IC fabrication process.

6.2 Describe Moore's law and how it explores semiconductor growth.

6.3 Describe about wafers.

6.4 Demonstrate different classifications of IC.

6.5 Give short notes on IC classification based on integration level.

6.6 Enumerate about wafer preparation process.

6.7 Define oxidation.

6.8 List out the type of oxidation process.

6.9 Differentiate dry oxidation and wet oxidation.

6.10 Examine the effect of the oxide thickness on oxidation.

6.11 List out the various oxidation technique.
6.12 Explain diffusion.
6.13 List out the type of printing mechanism.
6.14 Define Ion Implantation.
6.15 Enumerate about sputtering.
6.16 Differentiate analog and digital ICs.
6.17 Mention the advantages of the CVD method.
6.18 Give short notes on photolithography.
6.19 Explain the features of photoresist.
6.20 Point out the importance of metallization in IC fabrication.
6.21 List out the various types of IC packaging.
6.22 Summarize the different types of CMOS fabrication.

## Category II

6.1 Explain the process of oxidation.
6.2 Enumerate diffusion process and describe the diffusion atomic mechanisms.
6.3 Elucidate the process of CVD.
6.4 Discuss in brief about photolithography.
6.5 Give an overview of twin-tub process.
6.6 Illustrate the process of IC fabrication with suitable diagram.
6.7 Enumerate in detail about the CMOS inverter fabrication.
6.8 With neat sketch, elaborate ion implantation method in IC fabrication.

## KNOW MORE

## History

The first integrated circuit was invented by Jack Kilby in 1958. Making tiny ceramic substrates, sometimes known as micromodules, each containing a single miniature component, was an early concept for the integrated circuit. Then, components might be wired and combined into a compact bidimensional or tridimensional grid. Jack Kilby presented the US Army with this proposal, which at the time seemed quite promising, and it eventually resulted in the short-lived Micromodule Program. However, as the project gained momentum, Kilby developed a brand-new, ground breaking design: the IC. Kilby's IC was challenging to mass-produce since it featured external wire connections. The first monolithic integrated circuit was created in 1959 by Robert Noyce. Silicon was used to create the chip. In Noyce's monolithic IC, all the parts were housed on a single silicon chip and were connected by copper lines. Kilby's chip was built of germanium, while Noyce's design used silicon. Between 1961 and 1965, the Apollo Program at NASA was the single-largest purchaser of integrated circuits. Jack Kilby was awarded with the Nobel Prize in Physics on December 10, 2000.

**Jack Kilby**

**Robert Noyce**

## Real Time Application of IC

ICs are used in different forms. The real time application of IC includes:

- Speakers, microphones, and headphones
- Tablets and smartphones
- Wearables such as smartwatches
- Smart speakers like the Google Home or Amazon Echo
- Digital cameras, both still and moving
- Gaming systems and controllers
- Automobile infotainment systems
- LED lights or wall outlets for smart homes

## Interesting Facts about IC

The generation of technology and its switching device details and characteristics are tabulated in Table 6.3. For every generation the size and price of the semiconductor device decreased and its corresponding speed is increased. So, the recent ICs used in nano Bio-chips and AI have high speed, lower delay and lower power consumption.

**Table 6.3:** IC Generation and its Characteristics

<table>
<tr><th>Gen.</th><th>Duration</th><th>Switching device Used</th><th>Size (Decreases from 1st to 6th)</th><th>Speed (Increases from 1st to 6th)</th><th>Price (Decreases from 1st to 6th)</th></tr>
<tr><td>1st</td><td>1946-1954</td><td>Vacuum Tubes</td><td rowspan="6">Decreasing</td><td rowspan="6">Increasing</td><td rowspan="6">Decreasing</td></tr>
<tr><td>2nd</td><td>1955-1964</td><td>Transistors</td></tr>
<tr><td>3rd</td><td>1964-1977</td><td>Integrated Circuits</td></tr>
<tr><td>4th</td><td>1978-1982</td><td>VLSI or Microprocessor</td></tr>
<tr><td>5th</td><td>1982-present</td><td>ULSI</td></tr>
<tr><td>6th</td><td>Present-future</td><td>Nano Bio-chips, AI</td></tr>
</table>

The generation of storage device and its corresponding OS & programming languages used are tabulated in Table 6.4.

**Table 6.4:** IC Generation of Storage Device

| Gen. | Storage device used | OS used | Programming Languages |
|---|---|---|---|
| 1st | Magnetic drum | Batch OS | Machine Lang. (bits) |
| 2nd | Magnetic core | Multi Batch OS | Assembly Lang. |
| 3rd | Magnetic core | Real Time-sharing OS | FORTRAN, COBOL |
| 4th | Semiconductor memory, Winchestor disk | Time sharing OS, MS DOS, UNIX | FORTRON 77, PASCAL, ADA |
| 5th | Optical disk | Windows,LINUX GUI-LINUX, Android | High level language, C, C++, JAVA |
| 6th | Flash memory | Windows 10, cloud OS | High level language, Python |

## Logic Gates IC and its Semiconductor Manufacture Companies

A logic gate is a component of electronic equipment that executes logical processes. Electronic devices use seven different types of logic gates. They are NOT, OR, AND, NAND, NOR, EX-OR, and EX-NOR gates. There are basic gates, universal gates, and exclusive gates among the seven different types of gates. These come under the 7400 series integrated circuits.

- **Basic Gates:** Digital electronics uses three different types of basic gates. They are NOT, OR, and AND gates.
- **Universal Gates:** Digital electronics uses two types of universal gates. They are NAND and NOR gates.
- **Exclusive Gates:** Digital electronics uses two types of exclusive gates. They are EX-OR and EX-NOR gates.

The 7400 series has many logic gates integrated circuits (ICs). Table 6.5 shows the logic gate IC number with its corresponding logic gate name.

**Table 6.5:** 7400 Series Logic Gates IC

| S.No | IC Number | Logic Gate Name |
|---|---|---|
| 1 | 7400 | NAND |
| 2 | 7402 | NOR |
| 3 | 7404 | NOT |
| 4 | 7408 | AND |
| 5 | 7432 | OR |
| 6 | 7486 | EX-OR |
| 7 | 74266 | EX-NOR |

There is a two-part serial number on every IC chip. The manufacturer's information is indicated in the serial number's first segment. The technical details of the IC are listed in the second portion of the serial number. Similar IC chips with the same technical specifications are produced by many IC manufacturers. First, two or three-letter prefixes identify the device's manufacturer. Some examples of 7400 IC manufacturers and their corresponding IC prefix terms are tabulated in Table 6.6.

**Table 6.6:** 7400 Series IC Manufacturer and its Corresponding IC Prefix Term

| S.No | Prefix Term | IC Manufacturer Name |
|---|---|---|
| 1 | SN | Texas Instruments for commercial processing |
| 2 | SNV | Texas Instruments for military processing |
| 3 | MC | Motorola |
| 4 | M | ST Electronics |
| 5 | DM | National Semiconductor |
| 6 | MM (or) DM | Fairchild Semiconductor |
| 7 | UT | Cobham PLC |
| 8 | SG | Sylvania |

Next, two digits represent the temperature range. Some of the examples are listed as follows:

- 54–indicates the military temperature range
- 64–indicates the industrial temperature range
- 74–indicates the commercial temperature range
- Next, zero to four letters indicate the logic subfamily. Some of the examples are listed as follows:
- Zero letters–indicate the basic bipolar TTL
- C – indicates CMOS
- H – indicates high speed
- L – indicates low power
- S – indicates Schottky
- LS – indicates low power Schottky
- F – indicates fast
- HC – indicates high speed CMOS
- AHC – indicates advanced high-speed CMOS

Next, two or more digits are assigned at random to represent the device's function. Each family contains several hundred different devices. Examples are 7400, 7402, 7404 etc. The package type, quality grade, and other information may be indicated by additional suffix letters and digits; however, this varies greatly by manufacturer. These details can be obtained from the IC device datasheet.

## REFERENCES

1. Ankaj Gupta, Kumar Shubham, *Integrated Circuit Fabrication,* CRC Press, 2021.
2. Lih J.Chen, *Silicde Technology for Integrated Circuits,* IET Publication.
3. Coasts Spanos, Gary S. May, *Fundamentals of semiconductor manufacturing and process control,* John wiley and Sons Publication, 2006.

4. Peter Shepherd, *Integrated Circuit design, fabrication, and test,* McGraw-Hill Publication, 1996.
5. BOSE, *IC Fabrication Technology,* McGraw Hill Education Private Limited, 2014
6. *https://edu.ieee.org/eg-aiet/integrated-circuits-ic-technologies/*
7. *https://en.wikipedia.org/wiki/Integrated_circuit*

# TABLE OF PHYSICAL CONSTANTS

## Table I: Physical constants

| | |
|---|---|
| Avogadro's number | NA = $6.02 \times 10^{23}$ atoms per gram Molecular weight |
| Boltzmann's constant | $k = 1.38 \times 10^{-23}$ J/K<br>$= 8.62 \times 10^{-5}$ eV/K |
| Electron charge (magnitude) | $e = 1.60 \times 10^{-19}$C |
| Free electron rest mass | $m_0 = 9.11 \times 10^{-31}$ kg |
| Permeability of free space | $\mu_0 = 4\pi \times 10^{-7}$ H/m |
| Permittivity of free space | $\epsilon_0 = 8.85 \times 10^{-14}$ F/cm<br>$= 8.85 \times 10^{-12}$ F/m |
| Planck's constant | $h = 6.625 \times 10^{-34}$ J-s<br>$= 4.135 \times 10^{-15}$ eV-s<br>$h/2\pi = \hbar = 1.054 \times 10^{-34}$ J-s |
| Photon rest mass | $M = 1.67 \times 10^{-27}$ kg |
| Speed of light in vacuum | $c = 2.998 \times 10^{10}$ cm/s |
| Thermal voltage (T=300 K) | $V_t = \frac{KT}{2} = 0.0259V$<br>KT = 0.0259 eV |

## Table II: International system of units

| Quantity | Unit | Symbol | Dimension |
|---|---|---|---|
| Length | meter | m | |
| Mass | kilogram | kg | |
| Time | second | s or sec | |
| Temperature | kelvin | K | |
| Current | ampere | A | |
| Frequency | hertz | Hz | 1/s |
| Force | newton | N | $Kg\text{-}m/S^2$ |
| energy | joule | J | N-m |
| Conductance | siemens | S | |
| Inductance | henry | H | Wb/A |
| Magnetic flux density | tesla | T | $Wb/m^2$ |
| Resistance | ohm | Ω | V/A |
| Capacitance | farad | F | C/V |

| Quantity | Unit | Symbol | Dimension |
|---|---|---|---|
| Potential | volt | V | J/C |
| Power | watt | W | J/s |
| Magnetic flux | weber | Wb | V-s |
| Pressure | pascal | Pa | $N/m^2$ |
| Electric charge | coulomb | C | A-s |

## REFERENCES FOR FURTHER LEARNING


1. David.J.Griffiths, *Introduction to Quantum Mechanics,* Pearson Prentice Hall, 2005.
2. Chenming Calvin Hu, *Modern Semiconductor Devices for Integrated Circuits,* Pearson Prentice Hall, 2009.
3. S.O.KASAP, *Principle of Electronic Materials & Devices,* MC Graw Hill Education, 2018.
4. S.Datta, *Quantum Transport:Atom to Transistor,* Cambridge University Press, 2005.
5. Robert F. Pierret, *Semiconductor Device Fundamentals,* ADDISON-WESLEY Publishing Company, 1996.
6. Donald A. Neamen, *Semiconductor Physics and Devices Basic Principle,* MC Graw Hill Companies, 2012.
7. Ben G, Streetman, Sanjay Kumar Banerjee, *Solid State Electronic Devices,* Pearson Education Limited, 2016.
8. Ankaj Gupta, Kumar Shubham, *Integrated Circuit Fabrication,* CRC Press, 2021.
9. Lih J.Chen, *Silicde Technology for Integrated Circuits,* IET Publication.
10. Coasts Spanos, Gary S. May, *Fundamentals of semiconductor manufacturing and process control,* John wiley and Sons Publication, 2006.
11. Peter Shepherd, *Integrated Circuit design, fabrication, and test,* McGraw-Hill Publication, 1996.
12. BOSE, *IC Fabrication Technology,* McGraw Hill Education Private Limited, 2014
13. *https://edu.ieee.org/eg-aiet/integrated-circuits-ic-technologies/*
14. *https://en.wikipedia.org/wiki/Integrated_circuit*


## CO AND PO ATTAINMENT TABLE

Course outcomes (COs) for this course can be mapped with the programme outcomes (POs) after the completion of the course and a correlation can be made for the attainment of POs to analyze the gap. After proper analysis of the gap in the attainment of POs necessary measures can be taken to overcome the gaps.

**Table for CO and PO attainment**

| Course Outcomes | Attainment of Programme Outcomes<br>*(1- Weak Correlation; 2- Medium correlation; 3- Strong Correlation)* | | | | | | | | | | | |
|---|---|---|---|---|---|---|---|---|---|---|---|---|
| | PO-1 | PO-2 | PO-3 | PO-4 | PO-5 | PO-6 | PO-7 | PO-8 | PO-9 | PO-10 | PO-11 | PO-12 |
| CO-1 | | | | | | | | | | | | |
| CO-2 | | | | | | | | | | | | |
| CO-3 | | | | | | | | | | | | |
| CO-4 | | | | | | | | | | | | |
| CO-5 | | | | | | | | | | | | |
| CO-6 | | | | | | | | | | | | |

The data filled in the above table can be used for gap analysis.

# INDEX

## A



## B

## C

## D

## E

## J



## K



## L

## M



## N

## O



## P

## Q

## R



## S